Creole Son

Creole Son

An Adoptive Mother Untangles
Nature & Nurture

E. KAY TRIMBERGER

Introduction by Andrew Solomon

Louisiana State University Press

Baton Rouge

Published by Louisiana State University Press
Text copyright © 2020 by E. Kay Trimberger
Introduction © 2020 by Andrew Solomon
Afterword © 2020 by Marc Trimberger

DESIGNER: Mandy McDonald Scallan
TYPEFACES: Whtiman, text: Gotham, display
PRINTER AND BINDER: LSI

Names: Trimberger, Ellen Kay, 1940– author. | Solomon, Andrew, 1963–
author.
Title: Creole son : an adoptive mother untangles nature and nurture / E.
Kay Trimberger ; introduction by Andrew Solomon.
Description: Baton Rouge, LA : LSU Press, 2020. | Includes index.
Identifiers: LCCN 2019042892 (print) | LCCN 2019042893 (ebook) | ISBN
978-0-8071-7310-7 (paperback) | ISBN 978-0-8071-7324-4 (pdf) | ISBN
978-0-8071-7325-1 (epub)
Subjects: LCSH: Trimberger, Ellen Kay, 1940– | Trimberger, Marco, 1981– |
Interracial adoption—California. | Interracial adoption—Louisiana. |
Birthparents—Louisiana. | Families—Louisiana. | Adoptees—Drug
use—United States. | Nature and nurture—United States.
Classification: LCC HV875.64 .T75 2020 (print) | LCC HV875.64 (ebook) |
DDC 362.73092 [B]—dc23
LC record available at https://lccn.loc.gov/2019042892
LC ebook record available at https://lccn.loc.gov/2019042893

For my son and our extended families, biological and adoptive

Definition of a Louisiana Creole:

Gumbo, that's what we are . . . a little bit of everything, Black, White, French, Indian, a delicious stew.

—BLISS BROYARD, *One Drop*

CONTENTS

Photographs follow page 62.

INTRODUCTION

Andrew Solomon

This is both a rigorous and a brave volume, both a meticulous study of behavioral genetics and a deeply personal story of the complex relationship between the author and her adopted son, Marco. It explores cultural touchstones such as race, addiction, and love, and it does so with compassion and sadness. It argues for the primacy of nature over nurture, for the vanity of molding an adopted child so that he becomes a reflection of his adoptive parent. But Kay Trimberger's approach is never reductive. This is a book about the same lessons learned two ways: painfully, by living them; and restoratively, by studying them. Kay Trimberger is given to neither effusion nor self-pity, and her intellectual nature frames this book, but the emotions nonetheless run high. It is the story of her decision to adopt her Creole son, Marco; of his enchanting childhood; of his tumultuous adolescence; of his dysfunctional early adulthood; and of the crazy range of hopes and frustrations she endured through each of these phases as she strove for better consequences. It is also a warning about the dangers that ensue when public opinion goes too far toward narratives of exclusive parental power (as it did in the post-Freudian mid-twentieth century) or too far toward complete parental impotence (as it did in the early twenty-first century). "As a sociologist, I noticed that this attempt to make an adopted family identical to a biological one coincided with the decline of extended family ties and the idealization of the nuclear family," Kay Trimberger writes. She sets the social context for her book in lucid prose, painting a succession of hard-won insights. Like many parents, she loved her son more deeply than she knew him and she had to discover who he was as he emerged into a troubled coherence.

At the end of the seventeenth century, in his *Essay Concerning Human*

Understanding, John Locke wrote, "Let us then suppose the mind to be, as we say, white paper void of all characters, without any ideas. How comes it to be furnished? Whence comes it by that vast store which the busy and boundless fancy of man has painted on it with an almost endless variety? Whence has it all the materials of reason and knowledge? To this I answer, in one word, from EXPERIENCE." He wrote in a famed 1690 letter, "I imagine the minds of children, as easily turned, this or that way, as water itself." These reflections were made, it has since been wryly noted, by a man who never had children himself. Yet his supposition that children are *tabula rasa* has penetrated our culture, and a glut of parenting advice books effectively offer you a chance to shape your child however you like. Kay Trimberger maintains that nature is predominant but is profoundly shaped by nurture.

The Lockean idea of an all-powerful nurture reached an alarming apotheosis in the diverse work of the psychiatric pioneers Sigmund Freud, Karl Jung, Leo Kanner, Bruno Bettelheim, Margaret Mahler, and others who believed that a parent's actions could make an otherwise straight child gay, an otherwise sane child schizophrenic, an otherwise communicative child autistic. Those attributions were discounted in the later twentieth century, as clinicians observed that no parent is powerful enough to bring about such profound alterations of a child's core being. But the notion of what we might call parental determinism continues to surface today. Parents trade on the cachet of their children's accomplishments as though implicated in them, and there is seldom a news report of adolescent psychopathy that does not implicate parents or parental figures. Addiction is the result of an impoverished childhood, adult divorce the consequence of surviving divorce in the early years. Books of parental advice explain how to bring up a happy child, a successful child, a calm child, a child with any number of other attributes that are in many respects entirely outside of a parent's power. Locke's basic understanding of the strength of nurture has continued to inform the idea that parents can make their children into themselves.

The testing ground for such theories has been adoption. That is not because many researchers in these fields had any particular interest in adoption itself; it is because adoption gives the most arresting context for the separation of nature and nurture, as children are brought up by people to whom they have no genetic relationship, and whose predispositions they are therefore unlikely to have inherited. There have been many such studies. The most intense ones look at identical twins separated at birth and brought up without contact. But the studies that are simply of adoptive families have almost equal power. The nurture arguments support the practice of adoption's heroic narrative, in which a child who would otherwise have been neglected and alone is swept up into the loving arms of parents who can give that child everything. This was the dominant narrative during the Baby Scoop era that went from World War II until the 1970s. But there is an antiheroic narrative of adoption, too, one that has been on the rise for the last thirty years or so, in which people of relative privilege take children out of the only context that would ever have made sense to them and try to force them into norms that are antipathetic to them. For the time being, at least, that is the ascendant one. It acknowledges the force of multifaceted nurture complications that no one had identified to Kay when she adopted Marco, including so-called adoption trauma—the sense of rejection and displacement that many adopted people derive from the loss of birth parents, family, and context. Such early disruption can lead to difficulties with trust, attachment, and permanence.

Kay Trimberger adopted Marco toward the end of the Baby Scoop era with what then seemed an idealistic belief that a transracial and transclass adoption could unfold smoothly in the presence of sufficient goodwill, especially in a race- and class-diverse environment. That belief led to positions that Kay Trimberger herself has identified as naïve. She gave her son a childhood suffused with close attention, though she failed to discover or prevent an episode of possible early sexual abuse. But bringing up Marco affected Kay, and being brought up by Kay affected Marco. They both bear witness to the power of that love. She has continued to prop him up and try to help him move forward, while he has contemplated ideals that would surely have been outside his imagination without his early exposure to her world of intellect, decorum, and dependability. They may form an odd couple, but they form a couple nonetheless. I am a reader of their book, but I also know and like them both, and I find the tenacity of their bond deeply touching. In these pages, Kay has never glossed over the immense difficulties she and Marco faced. In many ways, Marco could never become middle-class, and Kay could never stop being middle-class. But neither has either of them ever given up on the other.

Kay Trimberger's investigation of behavioral genetics—she collates studies in Colorado, Iowa, Texas, Sweden, etcetera—turns around all that research that was coincidentally about adoption to apply to adoption itself. She indicates that adopted children resemble their adoptive parents during childhood and their birth parents by the late teen years, an idea that fits with the story she has to tell. This inclination toward biological origins continues to manifest itself into adulthood, when adoptees will have professional attainments closer to those of their birth parents than to those of their adoptive parents. Kay thought she was losing Marco to adoption trauma, but she came to understand that she was losing him in largest part to a biological destiny of addiction over which she had never had the slightest control. Had she known about the addiction history of both his birth father and birth mother, however, she might have been better equipped to recognize what was going wrong earlier and perhaps to intervene more successfully, to minimize the effects of this inheritance. *Creole Son* lays out what she has called "the roller-coaster of my fears and hopes"—the constant struggle and frustration inherent in dealing with anyone whose behavior is completely alien to yours. It is a struggle many non-addicts have with addicts, and the ride is often an alarming one. Whether Marco's dysfunction was not only about the elusive nature of sobriety but also about his life with Kay or his abandoning father was very difficult to know—difficult not only for Kay, but also for Marco.

"Like 87 percent of adoptive parents in a national survey, I would make the decision again to adopt, although knowing what I know now, I'd do many things differently," Kay writes. But she likewise observes that parents "need to ask how they can compromise between their needs and those of their adopted child." Kay never backs away from self-questioning: about the decision to adopt Marco; about her failure to protect him from early abuse; about the public school to which she switched him after giving him a more cloistered childhood; about the pressures she introduced to his life; about waiting to connect with his birth family; about the conflicts between how she wanted to live and what Marco appeared to need. Marco's family of origin, whom she met belatedly, are always treated with respect in this text. She went to extraordinary if belated attempts to bond with them, to understand the world her son came from. Her visits to Louisiana have an aura of pilgrimage. They are difficult but illuminating. She wants to like these people but she finds the time with them difficult, and her middle-aged whiteness is thrown into stark relief as she makes what can feel like obligatory conversation with them. It's partly race, partly class, partly discomfort with the huge role religion plays in their lives, and partly an inability to fathom the ubiquity of substance abuse among them. Yet she is also humbled by them, and acknowledges that the children Marco's birth mother raised had achievements beyond Marco's. She likewise recognizes that a real relationship with Marco's birth family would have had to be initiated in Marco's early childhood for the bonds to be fully realized. And she regrets that she didn't build up that deeper relationship earlier. "My biological family was his adoptive one; his biological family could have been my adoptive one," she writes with regret at the opportunity irretrievably lost. Ultimately, she celebrates Marco's having more family than she herself could provide, and she encourages his attachments in Louisiana. When his stepmother says, "Marco will always be my son," Kay does not experience loss. "Extending family ties, integrating biological, adoptive and chosen kin, brings nature and nurture together," Kay writes.

But addiction haunted Marco and his biological family. In the touching Afterword to this book, he writes, "'How great is this,' I remember thinking. 'I get to meet my birth family and I'm getting high at the same time.' Everything was just perfect. I couldn't believe how cool it seemed to have my cake and eat it too. But the reality was that there was nothing cool about it. Even though over time we all realized that this was not a good situation, we still kept it up." It was the complicated ambivalence around addiction that finally led Marco to cut many of his ties to his birth family, even as Kay was seeking out that birth family. But Marco retains a bond to a half-sister on his father's side. She was younger than he and leading a quite different life, but she was free of addiction and she, too, had grown up without a father.

Berkeley has always been at the forefront of progressive thought, a tendency that has sometimes been visionary and sometimes tragically adolescent. Kay Trimberger's book contains complex reflections on class, a set of divisions that she had grasped intellectually, but came to understand emotionally as they pertained to her life with Marco and with his biological family. She has said that she is grateful for the introduction to other worlds than her own, but she also demonstrates how much she had to do to be at ease in such other worlds, how frustrating she sometimes found it to be unable to read between the lines in a society so different from her own. She understood Marco's consistent attraction to the black people he perceived to be more like him, but she struggled with giving up on the values she had worked so valiantly, and so unsuccessfully, to instill in him. She wanted him to work, to support himself, to function in the professional marketplace. She wanted him to live up to the middle-class values relating to his intelligence and privilege, for him to give back to society according to what had been given to him. Instead, Marco became homeless, mired in drug use and accompanied by a girlfriend who was an addict with a personality disorder. Kay did not express her disapproval, but the gap between what she had imagined and the reality she confronted had become so vast that there was no need to articulate it.

Robert Frost said, "Home is the place where, when you have to go there, / They have to take you in." And Kay's house was in that sense home to Marco. They lived close to each other; some of the time, he lived in her house. But they lived in worlds between which there was very little common territory. Kay writes movingly of her pain and disappointment, of how she couldn't bear to visit the places where Marco resided because they were simply "too depressing." The struggle she faced daily was how to fulfill her role as a mother to a son she loved and yet preserve her own sanity and dignity. And she was ultimately able to do that, to find a place where Marco could heal, allowing her to heal, and introducing them to a new shared life, happier than before. Though its permanence is unclear, she did, finally, achieve something of a happy ending.

Creole Son

Prologue

M ine is not a story of how an adopted son finds his birth parents and turns his life around. Nor is it a horror story about adoption. Looking back from my thirty-fifth year as an adoptive mother, I can say it is a complex love story and a cautionary tale.

In 2006 when my son was twenty-six, I helped him reunite with his Louisiana birth families, one black/Creole and one white/Cajun. I did so because I hoped that a reunion might help him overcome his problems and become a responsible adult. That didn't happen. But these families' embrace of him and of me, and their openness about their lives, provided the spark that led to this book. Although Marco* and his birth parents were separated when he was five days old, and he was raised three thousand miles away in a different cultural and class setting, I was startled by the many unexpected similarities between them.

As a result, I set off on my own search for understanding, a search which led me to look into genetics, a topic that had never appealed to me and that I saw as leading only to dangerous social policies. I soon discovered the interdisciplinary field of behavioral genetics, which in the early 1970s separated from quantitative psychology and sought to distance itself from any connection with racist attempts at biological improvement. What especially caught my interest was that this field, in its attempt to disentangle the impact of nature and nurture, studied adoptive families.

---

*My son approved of using his real name. Unless I was given explicit permission, all other names in the book are fictitious. All are real people, and not composites, but I have sometimes changed details of their lives to protect their privacy.

Behavioral genetics analyzes the psychological and cognitive traits of adoptees over time and compares them to those of their birth mothers, adoptive parents, and birth and adopted siblings. Often these researchers use biological families as a comparison group. Behavioral genetics zeroes in on how genes and the environment interact and alter each other, together constructing individual behaviors.

I began to read behavioral genetics studies—not an easy task because of the use of scientific language and quantitative methods, some of which were unfamiliar to me. It was worth the work, for I gained a new understanding of my personal journey as an adoptive parent. I saw how Marco's development was influenced both by his genetic heritage from Louisiana and by the environment in which he grew up in Berkeley, California, toward the end of the twentieth century. Untangling nature and nurture led to new insights into how they work together, insights that are useful to biological families too. Readers will find some of my story most relevant to adoption, but more of it will be of interest to all families, especially those with addiction or other problems.

I tell a second, intertwined story—mine as a single mother seeking alternative family and community as I struggled to raise a black, biracial boy. Both my courage and my naiveté in adopting as a single mother mirror the times, the social rebellion and innovation of the 1960s and 1970s. Looking back, I analyze how some of this experimentation had a negative effect on Marco and me. But other experiences were more positive. As he made friends across class and race lines, I, to my benefit, was pushed beyond my previous experience. Our story reflects the personal pain, yet promise, too, involved in social change.

Finally, Marco's and my experience with his Louisiana families, where we each formed a lasting bond with a few members of the extended families, led me to envision a new model of extended family combining biological and adoptive kin. This model starts with, but goes beyond, current ideas about open adoption.

As I wrote this book, I struggled with the ethical and interpersonal concerns any parent faces when writing about her or his child. How would Marco feel about it? Talking with him, I found he approved of my telling *our* story of his adoption, and I gained the courage to write. But he didn't read any of the manuscript until after I had completed a second draft, several years later. I wondered how his reaction would affect our relationship. Would I publish if he objected? When he told me I had gotten the story right and that he gained personal insights from the science, I was pleased, especially when he accepted my offer to write an afterword. I am filled with pride at his openness, generosity, and lively prose. The book has led to a deepened bond between us.

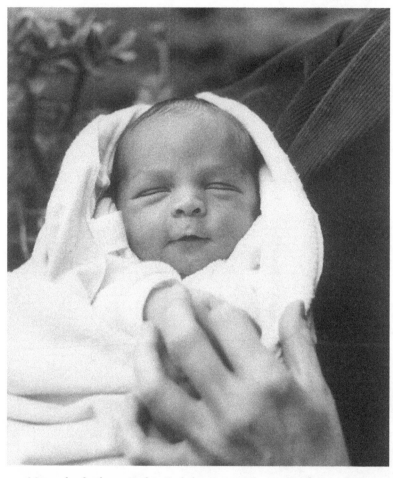

Marco the day he arrived in Berkeley, January 15, 1981, at five days old.
Photo from author's collection.

CHAPTER 1

How It All Began

O n the cold, sunny morning of January 15, 1981, Martin Luther King's birthday, I stood with anticipation at the San Francisco airport, waiting for my new son, a five-day-old mixed-race boy whom I named Marc, soon to be changed informally to Marco. A Caucasian doctor, who had said he wanted to "get the baby out of Louisiana," was flying him to me. Picking up a baby at an airport seemed surreal. But I wanted to be a mother, and as a forty-year-old single woman unable to conceive, I felt lucky to be adopting an infant.

Looking at photos of that day, I see a young and wide-eyed woman, trim in my jeans and red corduroy jacket, but looking solemn and scared. Everything had happened fast. I had heard about this baby only ten days earlier. No nine months for me. My housemates from our communal household in Berkeley had driven me to the airport with a donated baby car seat in the back, but the plane was late. Waiting at Gate 10, I turned and suddenly saw a tall, white man, formal in a sports jacket and red tie, and a petite woman, dressed in a black blazer and white blouse, walking down the hall toward us. Something she carried was wrapped in a blue blanket. It was the baby! A few minutes later, he was in my arms, with a light brown face topped by a fringe of black hair. Although holding him stiffly, I was full of joy. With his legs scrunched up as if he were still in utero, the compact baby, a strong six-and-a-half pounds, could almost hold up his head. I immediately fell in love.

We sat in the airport coffee shop to sign the papers, very straight-forward. Although the birth mother was white, the doctor told me in a strong southern accent that a black foster mother had had the baby for the past four days. She said that he was alert and easy and she was upset that we were taking the baby out of Louisiana. This information gave me qualms about out-of-state adoption. But then the doctor said of our black waitress: "Waitresses in Louisiana aren't so uppity." What had she said or done? I had no idea. For me, his statement validated the idea that bringing this child to California was a good idea. I rationalized that he would grow up in a less racist world.

Three years earlier, I began my trip that ended at the airport. Still uncoupled and past thirty-five, I realized I wanted to be a mother anyway. As a college and graduate student in the 1960s, exposed for the first time to progressive thought and action, I felt as if I were a caboose being dragged into history. But as an adult woman and a feminist in the 1970s, I was in the driver's seat. I could parent without a partner. Feminism, therapy, and inheriting my father's nurturing temperament contributed to my self-assurance that I could be a better mother psychologically than mine had been.

After I decided to try to become a single mother, adoption was only one of the paths that I pursued. I tried to get pregnant, but I was not a woman who yearned for the experience of pregnancy and I was skittish about childbirth. I also wondered how colleagues and students would react to a single, pregnant professor. In the late 1970s and early 1980s, women faculty were a minority, married women faculty were rare, pregnant married women faculty even rarer, and, in my experience, pregnant single women faculty were nonexistent. I felt embarrassed at the prospect of being single and pregnant in the classroom. Still, I longed to be a mother, and pregnancy was the easiest way.

I had a younger boyfriend whom I saw neither as a permanent partner nor as a possible father. He would soon be leaving Berkeley. So I secretly put a hole in my diaphragm. Also at this time, my male gynecologist gave me artificial insemination allegedly with semen from an early sperm bank, no questions asked and no preferences permitted. He also provided a fertility drug. When after six months I was not pregnant, I didn't investigate my infertility. The first test-tube baby was born in these years (1978), but in-vitro fertilization was not yet widely available.

I moved easily to the alternative of adoption, but I didn't think about it in any depth. I didn't ask myself then, didn't even consider the question, of what it might mean to adopt the baby of a stranger. Even though I was a sociology professor, interested in a range of historical, social, and psychological topics, my three-year journey to actually bring a child into my life did not lead me to research or to read about adoption. Becoming a mother was an emotional, not an intellectual, decision. Investigation was what I did for work; parenting would be something different. It would give me the opportunity to love and be loved. I imagined that loving a child would have none of the complications of my unsuccessful attempts to find a love partner.

When I came to the airport that morning, I carried with me a number of assumptions about adoption and child rearing. I assumed that I would not have any contact with the family of origin, since biology would have very limited impact on my son's development. Like other social scientists, feminists, and adoption experts of that era, I presumed nurture was everything, each infant a blank slate awaiting parental inscription, with genetic inheritance providing a minimal, if any, influence. This was the perspective I learned as a graduate student in the 1960s and perpetuated as a teacher and scholar in the 1970s and beyond.[1] Such cultural determinism was a reaction against earlier genetic determinism which was rightfully seen as racist.

When one applied this theory of the primacy of nurture to adoption, one healthy baby was as good as another; what mattered was how the baby was raised. When my son was an adult, he might want to meet the white woman who gave birth to him and meet his black birth father in order to clarify his identity, but that would have no effect on me. My son, whatever his race, would share my white, middle-class values, with their emphasis on the importance of education and work to the good life.

As a white, idealistic supporter of the civil rights movement, I believed that white families who adopted nonwhite children would help integrate society. I took for granted that cross-racial adoption would create few problems for the child. I believed that together we would fit into my multiracial neighborhood in Berkeley, a mixed-race, progressive city. I didn't think about the potential impact on my biracial child of my household, family, and friends—all of whom were white.

I adopted with one more personal assumption, shared not by large sectors of society, but by a smaller group creating countercultural institutions in the 1960s and 1970s. I thought I could be a better parent in a communal household than in a nuclear family or in a single-parent household. Four years before adopting, I had co-created a three-person communal household in Berkeley, with the express desire to parent in an alternative family. I came to this venture as the product of social forces that changed the lives of many middle-class, white women of my generation.

We came of age in the 1950s and lived our twenties and thirties in the tumultuous 1960s and 1970s. Our lives were full of contradictions. These were mine: I rejected my mother's experience as a college graduate who became a full-time housewife and mother, one where she had excellent domestic skills, but was not effective at psychologically nurturing me and my younger brother and sister. She seemed frustrated with her life and was always nagging us, especially my father and me. I learned some of her homemaking skills, especially cooking and home decor, and wanted a domestic life, but also a career. A professor father whom I resembled in looks and talents, and who was happier and more nurturing than my mother, led me to presume I would become a parent in my thirties after my career was settled. But I had few models of women with a career, or even full-time work, and most of those were single women who were portrayed in dreary terms. In college and graduate school, I was told indirectly by a boyfriend and by a graduate school recruiter that I had to choose either marriage or a career—a choice my father did not have to make. Later, this message was contradicted by the feminist movement that said we could have marriages or partnerships that equally combined work outside the home and family obligations. Yet I found many men didn't accept this new social vision, or did so in name only. Feminism didn't give me, someone disabled by my mother's inadequate nurturing, the personal tools or skills to negotiate an egalitarian relationship. Nor did feminism present a positive view of single life. Although I almost always had boyfriends in my twenties and thirties, these relationships all failed to be permanent. Either I or the man always exited.

But then the counterculture movement promoted a new family form, one especially prevalent in Berkeley, where I moved in my early thirties. A communal household would be a way to create a new type of family, one more like a traditional extended family, but based on friendship and not blood. This seemed to fit well with adoption, and I thought it would be easy to transition from a successful communal household to a communal family.

All these assumptions—about biology, race, and alternative family formation—proved to be naive or wrong, and they had a negative impact on my ability to be an effective parent. Looking back, I don't think that more knowledge would have altered my decision to adopt, but it would have provided some kind of map to the unknown territory I was entering.

Providing a home for a child already born made adoption seem a morally superior option. I thought that my relatively conservative family, as well as my liberal colleagues, would be more accepting of adoption than if I were single and pregnant.[2] My parents had approved the unorthodoxy of a single woman buying a house in co-tenancy with a couple, so I thought they would support my adopting a child. I was right, about my father at any rate, and he gave me a modest sum to help with the adoption. My mother still believed a woman should have a career or be a mother, but not both.

I didn't anticipate how hard it would be for a single woman to find a baby to adopt. A 1965 change in California law permitted a single person to adopt through state agencies, and in 1977 I applied. As part of the interview, the social worker showed me pictures of children and asked me to indicate those that appealed to me. I pointed to a picture of two young Chinese girls who were sisters. "You will never get them because they are too desirable," the social worker said. "As a single woman, you will only get a child who is hard to adopt, probably one with developmental difficulties."

I was taken aback. As a single parent, with a demanding job as a professor in a university fifty miles away from where I lived, I could not accept a special-needs child. I would, I said, consider a child of another race, and I would definitely accept a mixed-race child with one white birth parent. My thinking was that I could bond most easily if a child had one white parent. Now, I don't think this was true; skin color would not have affected my ability to love a child. I didn't think about how hard life might be for a mixed-race child who didn't fit into either racial niche.

During the next three years, I was never offered a child by a public agency. Was it because I was single? Too picky? Racist? I'll never know. Probably it had nothing to do with me. Because of the development of the birth control pill in the 1960s, the legalization of abortion in 1973, and the feminist campaign to legitimize single motherhood, there were fewer babies available for adoption by the late 1970s.[3] However, the big change at that time was that more black babies were released for adoption. Before the 1970s, single motherhood was accepted in the African American community and most babies of unwed mothers were raised within the extended black family. By 1980, new social mobility for a minority of educated blacks and the impact of drugs in poor black communities disrupted extended black families, leading to more adoptable black babies.

But the 1970s were also the years when the National Association of Black Social Workers condemned placing black children in white homes. They and other black nationalists argued that transracial adoption would undermine the African American community. Transracial adoption, they claimed, would derail cultural identity and curtail the teaching of survival skills to black children in a racially oppressive society.[4] These views were and still are controversial within the black community, with some black intellectuals seeing them as a form of racial discrimination. The latter, along with others, supported a law passed by Congress in 1996 which prohibited any consideration of race in adoption for agencies receiving federal funds.[5] I was aware of these debates, which reinforced my insecurities about raising a black child and my preference for a mixed-race baby born to a white mother.

Because the possibility of adopting through a state agency seemed increasingly remote, I looked at foreign adoption. I discovered this option was just as difficult for a single woman. In the 1970s, none of the agencies for overseas adoption accepted a single parent, although I could try to organize a private foreign adoption. After a ten-day visit to Mexico, I discovered that I would have had to stay there for at least six months, hire a lawyer to deal with the bureaucracy, and probably bribe officials—a distasteful prospect.

I had heard earlier about a road to private adoption in the U.S.—hiring a lawyer or consultant to help compose and send letters to gynecologists in small towns in predominantly Catholic areas. Since Catholics were opposed to abortion, there would likely be more babies available for adoption there. You wrote a story about yourself and hoped that a young, pregnant woman would choose you. In 1978, a lawyer told me this procedure would be a waste of my time because a single woman would never be chosen.

Discouraged, I stopped the process. Was I putting my life on hold while I went on a fantasy quest? I knew no other single woman who was deliberately trying to become a mother either through adoption or artificial insemination.

In 1980, the year I turned forty, I decided to try again. Two factors encouraged me. I was awarded a fellowship from the National Endowment for the Humanities for research. I didn't have to commute or teach for fifteen months. At the same time, I heard about a single woman who was able to adopt using the new process of private adoption. Maybe what the lawyer had told me two years earlier was wrong.

Maybe a young single woman would choose an older single woman to adopt her baby. I found out the names of a few locals who had adopted privately and made phone calls to them asking for advice. I also made an appointment with a different lawyer.

Before my meeting with this lawyer, I received a phone call on January 2, 1981, from a married woman in Berkeley who had adopted two children through a private lawyer in San Francisco. This lawyer knew about a baby in Louisiana who would be available for adoption. In a small town in southwest Louisiana, a seventeen-year-old white Catholic girl of Cajun descent was pregnant. She was the youngest of a big working-class family and all of her siblings had college educations. One of her brothers had a Ph.D. The birth mother had received gynecological care since the third month and had decided to put the baby up for adoption.

Presuming that the baby was white, the gynecologist was going to give this baby to his friends. But in the eighth month, the birth mother told her parents and the doctor that the father was a Creole of color.

Despite Louisiana's extensive history of racial intermixing, this white doctor decided that it would be best to send this mixed-race baby out of state. He remembered a letter in his file from a Berkeley couple who wanted to adopt and called them. They were no longer interested, but they told their lawyer about the baby and he called another woman. She had completed her family and remembered talking to me. She rang me and told me how to contact the lawyer. Serendipity is indeed a large factor in adoption.

I liked the idea of adopting from Louisiana. I was attracted because a colleague from there had told me about its unique mixed-race history and culture. I called the lawyer, an elderly man with a huge head of white hair, who, I learned, was the most experienced private adoption lawyer in San Francisco. In a little bio I wrote about myself to send to the birth mother's family, I emphasized that I too came from a Catholic background and had two nuns in my family. I hoped that this would help them overlook my singleness.

By January 5, the family agreed to give me the not-yet-born baby under the following conditions: I would pay the birth mother's medical expenses of $4,000, give the lawyer a small fee, and cover the airfare for the Louisiana doctor to bring the baby to San Francisco. Later, I heard that he and his wife wanted a holiday in Carmel. With a little help from my father, I could afford these fees.

A baby boy was born on January 10, and by January 15 he was in my arms. After three years of a frustrating process, in two weeks I had become a mother. That morning at the San Francisco airport I started on a journey with a powerful desire to be a parent, but with no experience of doing so in an alternative family and hobbled by a tangle of misleading mythologies.

♦ ♦ ♦ ♦ ♦

On another January day, this time in 2010, Marco, at twenty-nine, drove a rental car with his black birth father, Reggie, sitting beside him. I sat in the back seat. To a casual observer, the two handsome light-skinned black men might be mistaken for brothers. Marco's father, despite a hard life, looked younger than his fifty-four years. But no one hearing the younger man call the other *Pops* or *Dad* would be surprised to learn that they were father and son. What would confound the bystander would be to overhear the driver address the much older, very white woman in the back as *Mom*. Here in Louisiana I felt uncomfortable, something I rarely experienced in driving with Marco in California.

Both men were five-feet-ten and both wore dreadlocks. Marco's hung long and thick, held back neatly in a ponytail by a twist of his own hair. Reggie's dreads were lighter in color, shorter, and thin—his hair damaged, he said, by lye thrown on him when he was in prison. Reggie was clean-shaven, while Marco sported a thin moustache that circled his mouth and connected with a short beard, nicely trimmed. Both men had the same North African nose, high forehead, large eyes, and prominent eyebrows, Marco's a bit thicker. Both were attractive, and they turned heads. I had always liked having a handsome son.

I looked nothing like them. Forty when I adopted Marco, now at sixty-nine I was showing my age. My pale, freckled skin and auburn hair tinged with gray contrasted with their darker skin tone and black hair. My large frame was different from their wiry bodies.

Continuing to mull over my observations, I acknowledged the similarities between Marco and his father, but I saw differences too. Reggie seemed shy, quieter than I had imagined, not as extroverted and verbally articulate as Marco. Maybe he was especially this way around me, but even as he began to relax, I saw that he lacked his son's personal charm. Marco also displayed a distinct manner of dress, having developed his own sense of style from the time he was in elementary school. At age six, he wore a fancy printed baseball cap turned backward. Today, he sported nicely fitting black jeans and a deep maize cotton-padded shirt over a black thermal, while Reggie wore baggy, gray sweats and a baseball cap. In my casual jeans and sweater, I looked more like Reggie.

Driving west from New Orleans on Interstate 10, the flat highway suspended above flooded land, we were on our way to visit Marco's white birth mother, Louise, her family, and more of Reggie's kin. Through the rain and fog, I saw bare trees and stumps covered in dripping, gray moss sticking out of the swamp. The dreamlike quality of the landscape, so different from the dry hills and vegetation of California, added to my feeling of unreality. Although Marco had known these families for three years, this was my first trip to Louisiana and my first time to meet them.

Over the years, I had daydreamed about going with Marco to meet his birth parents in Louisiana, but the people in my fantasies had been ghosts without features or personalities. If Louisiana and birth families were ghostlike to me, how much more confusing and dreamlike must they have been for him as a child.[6]

As we drove along the highway above the swamp, which I later learned was the Atchafalaya, the largest swamp in the U.S., Marco and Reggie began to talk about roadkill.

"Mom, do you know that Louisiana has more than any other state?" Marco asked rhetorically. I didn't, but before I could answer, Marco pointed to the left side of the road.
"There's a coon," he exclaimed.

Soon Reggie joined in. "Armadillo," he announced, and later, "possum, coyote, and muskrat."

They delighted in each sighting. I tried to follow, but from the back seat I saw little as we whizzed by each carcass. I was more interested in Reggie's terse stories of catching an alligator, making his own rabbits' feet key chains, and finding turtles to cook. Marco, too, was engrossed by these tales.

"I kill and shell the turtles," Reggie said, not giving us details of how he did that. "I then make a stew by cooking the meat very slowly, sometimes overnight, until it falls off the bone."[7]

As Reggie and Marco chatted in the front seat, I pondered my complicated feeling about the hunting skills which ran in both the black and white sides of Marco's Louisiana families, skills quite different from those to which Marco had been exposed in my urban culture and a sport of which I disapproved. After his first trip two years before, Marco had told me about his fascination with his uncle Don, his birth mother's older brother, a social worker, who often shot squirrels on his day off. He would cook some and freeze the rest to eat later. Don had explained to Marco that when their father was disabled by an accident on the oil rigs, he shot squirrels for the family's survival; now Don considered squirrel a delicacy. Marco detailed the cooking method for me. I found all this talk about roadkill and hunting fascinating, even though it was outside my urban world, but I couldn't envision adding squirrel to my recipe files.

This talk triggered a memory in which Marco at age five declared, "I want to learn how to fish." I had never fished and had no interest in learning. Unappealing to me was the knowledge of what hooks and bait to use for different fishing conditions, the hand coordination needed to tie on weights and hooks and bait, the patience to wait around for a fish to bite, and then the frenzy required to land it. I would have objected if Marco had wanted a gun to learn how to hunt, but fishing to me was benign and I wanted to support his interests. I found a male friend to help us buy supplies and teach Marco the fundamentals. He learned quickly. Afterward, I'd take Marco to a nearby lake or the bay where a gaggle of fishermen lined the shore. Even if no other kids were fishing, he gamely joined the fishermen, watching and asking for advice. I sat nearby in my beach chair, happily engaged in a book, wandering over to express pride when he caught one. By age eight, Marco, now an accomplished fisherman who cleaned the fish and prepared them for me to cook, gave free lessons to his friends and cousins. I was proud of him.

Eight was also the age when Marco started asking questions about Louisiana. So I introduced him to a colleague of mine, Velma, an African American woman whose family was from southern Louisiana, leading her to take a special interest in Marco. From her, Marco and I learned about contemporary Louisiana Creoles, a distinct mixed-race group with African, French, Spanish, and Native American roots. They are primarily Catholic, and older Creoles speak a French dialect unlike any other in Louisiana. Creoles are also known for their unique cuisine and music.[8] Later, I learned more about how Creoles in southwest Louisiana formed a distinct community. Bliss Broyard, in her biography of her father, the New York intellectual Anatole Broyard, who left his Louisiana Creole community and passed for white, says:

> [Louisiana Creoles] shared family recipes, holiday traditions, a love for storytelling. They attended the same high schools and churches. And they were united by a complex kinship in a community that, for better or worse, had been culturally, geographically, and legally segregated from both whites and African-Americans for more than a century and a half.[9]

Velma called Marco, like herself, *Creole,* and she introduced us to the culture's famous gumbo and to zydeco music. She said that her relatives in Louisiana considered themselves superior to African Americans. In California, she found that *Creole* had no meaning and she identified as African American.

Before Marco and I met Velma's extended family, now living in northern California, she had told me about the range of skin colors in her family, from medium black to white with blond hair. We observed this racial diversity some years later when she invited Marco and me to join her extended family at a zydeco concert and another time, for a Thanksgiving dinner. I loved exposing Marco to these cultural roots, which enriched my life too.

Adoption had robbed Marco of any direct learning of his cultural heritage and our contact with my friend was limited, since she lived fifty miles away from us and her children were grown. But from his first trip to Louisiana, Marco immersed himself in Creole traditions, especially the food. Marco's interest in food could have come from my love of cooking and the diverse cuisines to which I exposed him. But his preference for spicy food was not mine. From reading *Origins: How the Nine Months before Birth Shape the Rest of Our Lives,* by journalist Annie Murphy Paul, I learned that scientists have discovered that the foods a mother eats influence the fetus. Paul cites studies of amniotic fluid in which flavors as diverse as garlic, cumin, curry, fenugreek, mint, and vanilla have been detected. She concludes that a pregnant woman is educating her offspring in the flavor principles of her culture.[10] Paul's analysis made me believe that Marco had gotten his tastes from his nine months inside his mother and retained them all his life.

During Marco's first visit to Louisiana a few years before this visit with me, he had learned that he was Creole not just because of his mixed race, but because of a cultural heritage. Reggie grew up speaking French with a beloved Creole, Catholic grandfather. Marco's birth mother, Louise, is of Cajun descent through her mother, whose first language was also French. As a Cajun, Louise is a direct descendant of the French Catholic settlers who were expelled in 1755 from Nova Scotia, people who eventually found their way to southern Louisiana. Today, the Louisiana Cajun and Creole cultures overlap in language, food, and music. I learned too that Germans—my ethnic heritage—had settled in Louisiana and introduced the accordion, which became central to zydeco and other Louisiana musical traditions.

Perhaps this early exposure to Louisiana culture through Velma had helped Marco develop a Creole identity, which became fully realized when he finally reached Louisiana and engaged this heritage first-hand. During my visit, he introduced me with pride to boiled crawfish, shrimp étouffée (a sauce served over rice), jambalaya (rice mixed with chicken and meat or with a mixture of shrimp and other seafood), boudin (a Louisiana sausage), and oyster po-boy sandwiches. I enjoyed the flavorful food, even though some of it was a little too spicy for me, and I was pleased that Marco and his Louisiana family enjoyed food as much as did my family and I.

When he was eight, I had tried another avenue to introduce Marco to Louisiana culture. I borrowed some videos about southern Louisiana by documentary filmmaker Les Blank. We watched them together. They were too arty for a child and I don't think Marco got much out of them, but they were tantalizing to me. I was overwhelmed by the images of water and fishing. Had I read then what I have now, I would have gotten a hint that Marco's love of fishing might have something to do with genetic factors beyond my influence.

Driving along, I mulled over what Reggie's wife of four years, Jeanette, had told me the night before. I had sat talking with Jeanette, a large, pretty, and friendly black woman in her late thirties, in the living room of their small two-story townhouse in a suburb of New Orleans. Here Jeanette had been able to continue her career as a city employee after fleeing New Orleans with her young son during Hurricane Katrina. Jeanette reminded me that until three years ago, Reggie hadn't known that his son Marco existed, and Marco had grown up three thousand miles away in very different circumstances.

"But their personal habits are so much alike!" Jeanette exclaimed. Reggie had trouble keeping a job and often felt anxious in public, so Jeanette was the primary breadwinner, leaving Reggie responsible for the cooking and for keeping the house clean. Long before meeting Reggie,

Marco expressed the wish that he could be a house husband, as he too felt anxious outside the house. Like his father, he was competent and hardworking but found it hard to take initiative in public. Since they had no contact, Marco could not have learned this behavior from his father. These facts challenged my belief that genetic heritage was unimportant.

CHAPTER 2

Searches

During his middle childhood, Marco and I had talked about his birth parents and I had shared what little information I had obtained about them through the private lawyer. Adoption professionals in the 1980s and into the 1990s recommended reunions with birth parents only after the adoptee was an adult, usually over age twenty-one.[1] So I told Marco that he could contact his birth families when he became a grown-up. I assumed that after Marco became an adult, he might want to meet them, a process that he would initiate and would not involve me. But by age twenty-five, Marco had not become a self-supporting adult, and he was too frightened of rejection to pursue a reunion.

He was enthusiastic, however, when I reported that I had met a woman at a book talk who told me how to contact a finder, someone who for $200 would do an internet search and then make the first call to the birth parent. Later, I learned that two-thirds of adoption reunions are facilitated by an intermediary.[2]

Marco especially wanted to communicate with his black birth father, something I had anticipated many years before.[3] When he was four years old, I had the foresight to ask the San Francisco lawyer who had arranged the adoption to obtain the birth father's name. But more than twenty years later, the finder couldn't locate him with that name. Within a day, however, the finder was on the phone with Louise, Marco's birth mother, who had long hoped for this call. Louise told the finder that she was eager to talk to Marco and to introduce him to her family, which included two more mixed-race children.[4] She would help him find Reggie.

The next day, with the finder mediating, Marco and Louise spoke on the phone for thirty minutes. She wanted them to get together soon. Though this seemed like an ideal reunion story—no secrecy, no ambivalence, and complete acceptance—Marco had never liked change or upheaval. He needed time. It took him six months before he accepted Louise's invitation to visit Louisiana. But after the first ten-day visit, he wanted to return for a longer stay.

I was thrilled when the families of both his white mother and his black father welcomed him. It's uncommon for an adoptee to find two birth parents, rare for them both to be open to a reunion, and even more unusual for an adoptee to spend one six-month period and then another four months living with a birth parent and becoming acquainted with relatives. Given that he had grown up in Berkeley with me, a white professor, I was surprised at Marco's comfort with these relatives from a different class and culture.[5]

It is often assumed that adoptive parents will feel threatened when the adoptee reunites with birth parents. I had no such hesitation, because I was looking for answers to my concern about Marco's failure to create a stable life for himself, despite a happy and well-adjusted middle childhood, where he did well in school. But in eighth grade, he started to change and his grades plummeted. High school was even more problematic as he attended four different schools without graduating (although he obtained his high school equivalency degree). He was arrested at age seventeen for selling marijuana and in his twenties for drug possession and driving without a license. After high school he was dismissed from several retail jobs and dropped out of community college.

He was always losing things—a passport, a driver's license, several I.D. cards, innumerable phones, a storage locker full of furniture, clothes, childhood pictures, and his grandfather's carved African animals. He also lost a used Toyota I had bought him that had clocked only seventy thousand miles. When he was stopped for driving without a license, the car was confiscated. I was angry and refused to pay to get the car back. At age twenty-five, he was living on a friend's couch and together they sold scrap metal, did handyman jobs, and sold marijuana. This was not the life I had anticipated for him.

I had become used to Marco's frequent misdemeanor charges, most of which meant only a night or two in jail and a fine. He knew how to treat police with respect and they usually responded with leniency. Once he had gone in for a ten-day jail sentence, but he was released after five days. "The jail was too crowded," he said, "and they liked me." Eventually, I stopped paying his fines, so often he had outstanding warrants. He never seemed to get arrested for these, however. Unmet probation requirements and fines seemingly evaporated from the computer system, perhaps because they were minor offenses or because of statutes of limitation. I had mixed feelings about this. Jail time, I believed, was not a redeeming experience. But shouldn't his actions carry some consequences? I was relieved, however, that he had never had a DUI and had no felony charges.

Only when Marco was in his mid-twenties did I admit that he had a substance abuse problem, and I could sometimes use the word *addict*. I finally saw that drug and alcohol use could largely account for his delinquencies. It took me so long, I think, because, as my friends said, I was a substance abuse virgin. I never smoked, drank little, and controlled my attraction to sweets. Beyond a Wisconsin cousin, no one in my extended family had problems with addiction. When I tried marijuana, I choked and could not inhale. I remembered that when my professor boyfriend took me to Morocco with his millionaire friend and his girlfriend, they all made fun of me because I couldn't join them in smoking hashish.

I lived with this professor in my early thirties for more than two years. I thought he was my soul mate, but in reality he was fifteen years older and my mentor. I found him moody and unpleasant, and he treated me badly. I thought depression explained his actions. Not until he socked me in the eye, and I landed in the emergency room, did I leave. Although there was no permanent damage, I was humiliated that I had stayed so long. Unlike some battered women, I had a teaching job, an income, and friends.

After I left, one of his departmental secretaries said to me, "How could you have lived with that alcoholic?" This was the first, but not the last, time that I failed to recognize addiction in an intimate relationship. Ironically, my adopted son as a young man looked a bit like the professor, with the same dark good looks and thin body, and, as it turned out, a similar predilection to addiction. But he never hit me or, to my knowledge, any other woman.

Believing as I did then that parents were the major force in a child's development, I didn't understand how Marco could have become an addict. When he was eight or nine, he regularly reprimanded a visiting friend from New York who smoked on the back deck, saying it was unhealthy. Was it my fault that he now smoked and drank and used drugs? Where had I gone wrong? I was too ashamed to seek help, only able to talk to some of my friends. They urged me to go to Alanon, a national support group for family and friends of alcoholics and addicts. But only after months of hesitation did I follow their advice.

Meanwhile, I began to read what adoption scholars and researchers had to say. Their work indicated that adoption itself might be the source of his need to use substances. The dominant adoption theory of that time, and one still prevalent, emphasizes the psychological impact of adoption on a child. This theory postulates that adoptees, even those placed as infants, suffer a deep psychological loss, one that also involves a displacement of ethnic and genealogical history. This loss, they say, leads to a weak sense of self and low self-esteem.[6]

At the time, this theory seemed to fit Marco. As a small child, he always deferred to friends in deciding what to eat or play. He deferred to me as well. Only in clothes did he develop his own style. This psychological theory told me that Marco's weak sense of self could have led to substance abuse, perhaps as self-medication, but the mechanism was unclear. Certainly, I knew that not all adoptees went down this path. "Why him?" I wondered. Perhaps his reunion with his birth families would shed some light on my questions.

During his first ten-day trip to Louisiana, Marco met his birth mother and father, half-siblings, and aunts and uncles on both the Cajun and Creole sides of his family. Upon his return, he bragged: "My visit was the most interesting thing that has happened to those families in a long time." Basking in the glow of reunion, he made a picture necklace fashioned out of plastic holders, the kind that convention attendees or hospital employees wear to hold a name tag. He cut and reassembled five of them, linking them together with paper clips. Marco put pictures of his newly discovered relatives back to back in each holder so he could show off as many of them as possible. Around his neck, they twirled and sparkled like a string of jewels. The necklace expressed his joy at finding this family. He told me about each of his relatives pictured there, especially his birth mother and father. As a result, I already knew a lot about them before we met.

In Marco's necklace, Louise, his birth mother, at age forty-four was an attractive, short white woman with her black hair cut short. In her photo, I saw she had Marco's heart-shaped face and large eyes, but otherwise bore little physical resemblance to him. Later, when I met her, I found that her extroverted and open personality matched Marco's. During his visit, Marco had learned the outlines of her life, which he related to me. The youngest of four children, Louise grew up in a working-class family whose economic status deteriorated over time, so that she came of age in a different situation than that of her much older siblings. Her father worked on the oil rigs, which paid well, but in his forties, when Louise was still young, an accident sent him into early disability retirement. After that, he hunted deer and squirrels and trapped frogs and turtles to eat and to sell. He needed to supplement his meager pension and what his wife earned as a store clerk.

I was curious about these people whom I had fantasized about all these years, so I encouraged Marco to tell me lots of details. He seemed comfortable doing so. I learned from Marco that Louise at seventeen had no say over the future of her child. Her mother and her much older sister arranged the adoption. Louise never saw the baby, never held or nursed him. Nor had she seen the picture of me that I had sent to her gynecologist. She knew that the baby was adopted in California, but she hadn't known that I was a single mother.

In phone conversations before Marco's first visit to Louisiana, Louise revealed to him that his birth father, Reggie, didn't know he existed. After giving the baby up for adoption, she had defied her family and gone off with Reggie to Texas, telling him the baby had died in childbirth. Since we know that African American families at this time were more accepting of babies born out of wedlock,[7] she might have feared that his family would want her to keep the baby. Louise was probably under extreme pressure from her white family to give up the baby, and she may not have wanted to have contradictory pressure from Reggie. Like many young women in her situation in those years, she may not have known what she herself wanted.

Louise and Reggie soon broke up. If Marco knew more details about the breakup, he did not tell me. Louise then married a black man in Texas, an adoptee raised by white parents.

"Marco," I said, "I don't think it was an accident that she married a black adoptee raised in a white family. She married someone like you."

"Yes," he replied. "I guess she was more upset about giving me up than she could say."

In her marriage, Louise gave birth to two children, the older boy, Aaron, less than two years younger than Marco. But her husband turned out to be a crack addict and Louise soon became addicted, too. She left him when her daughter was a baby, living for a short while on welfare in a homeless shelter. Her extended family in Louisiana urged her to come home, and they helped her buy a trailer. Louise never saw Reggie or her husband again.

When I first heard that Louise lived in a trailer, my mind immediately conjured up images of "white trash." Based on stereotyped images, I imagined a disheveled trailer park with unemployed and addicted residents, their unkempt children running wild. Later, when I visited, I found a large and spacious trailer house with three bedrooms, sitting, not in a trailer park, but alone on a country road, parked on her sister's property. Family connections led to a job in a Target store, and her mother and sisters helped with the children, who lived with strict rules.

These details, which Marco divulged in a matter-of-fact manner, left me feeling conflicted. I was heartened to know that Louise's white extended family accepted her back with her mixed-race children. I appreciated that ten years earlier she had told her other children about Marco's existence and that now she wanted them to get to know each other. My phone conversations with her before Marco's visit had been positive. I was impressed with how open, warm, and self-confident she seemed. I hadn't expected a woman with only a high school education to have such a strong sense of self. I became aware that I had class biases, which contradicted my image of myself as a liberal with sympathies across race and class.[8] But Louise's drug addiction history alarmed me. Could she have passed on the propensity to addiction to Marco?

The most important photo in Marco's necklace was that of Reggie. He was a light-skinned black and he was a man and they looked alike. Reggie, it turned, out, was a jewel with imperfections that it would take Marco some time to discover.

Before the visit, Louise warned Marco that although she had not seen Reggie for twenty-six years, she had heard gossip about his troubled life, one that involved jail and drug use. She knew he still had family in their hometown, an hour's drive away, and through them hoped to find Reggie. Soon after Marco arrived, Louise took a day off work and they drove there together. She went directly to a store where she knew one of his brothers worked. When Louise told the brother the truth about the adoption, he came out to the car to meet Marco. The brother's first words on seeing Marco were, "You are Reggie's son; you look just like him!" The brother told them that Reggie had married recently and was living in a nearby city.

Louise called him, and within a day Reggie showed up at Louise's trailer in Lake Charles. When Reggie arrived, he seemed pleased to meet his *new* son and to see Louise again. Reggie revealed to Marco and Louise that he had been in and out of jail for twenty-five years on drug and petty theft charges. Marco told me later that Reggie survived prison life by drawing and writing poetry. Like Marco, Reggie seemed to be nonviolent and artistic. Like the father he had never met, Marco had served jail sentences, although fewer and shorter ones. Upon hearing this story, I wanted to investigate how they could share so many characteristics.

When I had asked Marco how he differed from his father, he replied, "Not a heck of a lot, but I have better tools." These tools—his socially adept personality and his social class advantages— along with the California penal system, which was more lenient than that in Louisiana, had saved Marco from Reggie's longer prison terms.[9]

Back in Berkeley, Marco was constantly on the phone with Reggie, sometimes three or four times a day, obsessed with a desire to return to

Louisiana and plan a life with his new father. Marco fantasized about how he and Reggie could live and work together.

"Mom," Marco cajoled, "for $7,500, you could buy the house where I was conceived, the small place where Reggie lived as a teenager with his grandfather."

I was skeptical, since he had told me that the house had been uninhabited for more than ten years and was in danger of being demolished. But Marco continued, "We could fix up the house, live there, and start a handyman business together."

"What about Reggie's wife and her son?" I asked. "This house is more than one hundred miles from where they now live and where Jeanette has a good job." Marco didn't have an answer.

Both Louise and I pointed out a number of other problems with this plan. They would need a truck. But Reggie didn't have a driver's license; nor did Marco since his had been taken away for fishing without a license. Marco never paid the fine. Neither of them had any business experience or showed any aptitude for it. Most important, Marco now admitted to me that Reggie was a crack addict. Marco vowed that he could cure Reggie, who, inspired by the joy of having a son, would want to be clean and sober. I doubted this. Wouldn't Marco be more likely to use drugs with Reggie in order to forge a bond with his new father?

While Marco was in the honeymoon stage often described in adoption reunions, I was in shock at the discovery of two birth parents with addiction histories. I decided to voice my concerns to Louise, glad that we had bonded enough on the phone for me to call her.

For the first time, I brought up Marco's addiction. She already knew. During their visit, she told me, she had asked him what he used. This was a surprise to me. She told me of her own past use, but asserted that she had been clean and sober for seventeen years. She obviously knew a lot more about addiction than I did. She also concurred that Reggie was still addicted to drugs.

But Louise urged me to help Marco return to Louisiana to live with her for a while. She seemed to want to incorporate this new son into her family. "I conquered drugs on my own," she said, "and neither of my kids has a problem; I will watch Marco to be sure he doesn't use." Louise said that if I would pay Marco's bus fare, she would support him for a while and help him find a job. She vowed to keep him away from Reggie's influence.

Relieved to have someone else say that they would help with my son, I gave in to the plan. Since I felt that I was losing my son to the underground and drug culture here at home, I didn't feel threatened by the possibility of losing him to Louisiana. My greatest fear was that addiction would ruin his relationships in Louisiana, causing these relatives to reject him. Right then, however, I was happy to be relieved of a burden. I rationalized that living with his birth mother in Louisiana would be an important growth experience for him. I imagined that he would develop a stronger sense of self. I conveniently forgot a warning from a therapist I saw who said that finding his birth family would not cure Marco's addiction. So in less than a month after his first trip to Louisiana, I bought him a bus ticket, gave him fifty dollars and some food for the trip, and sent him back. Even after he left, I was still focused on understanding my adoption experience.

◆ ◆ ◆ ◆ ◆

The uncanny connections that we discovered in Louisiana between Marco and his birth parents sent me on a search of my own, a quest to understand genetics and how they might impact the lives of adoptive families. Secretly, I hoped that I would find genetic determinism to escape the "mother blame" of the "nurture was everything" theory, so that I could dispose of my parental guilt.

As a social scientist with little biological education, I began by looking at science journalism, but I soon turned to original research. I discovered that genetics alone could not explain Marco's positive behavior nor his problems; genes provide only probabilistic tendencies, not predetermined programming. In only a few cases, like Huntington's disease and cystic fibrosis, will alterations in one gene, or a specific set of genes, determine who gets the disease. Personality and cognitive traits, along with behavior and most diseases, depend on a complex set of genetic pathways, few yet identified. Moreover, all genes can be altered by the environment. Thus, genes provide probabilities for behavior and risk factors for disease, but do not indicate whether any individual will develop a specific behavior or succumb to a particular mental or physical disorder. Even so, I found plenty of evidence that babies do not come into the world as blank slates, as I had believed when I started my adoption journey.[10]

During my research, I was surprised to find the field of behavioral genetics, which I'd never heard of, one which originated in psychology but in 1970 had defined itself as a separate interdisciplinary field. Behavioral geneticists at this time were trained primarily as psychologists and psychiatrists with skills in quantitative statistics.[11] What surprised me even more was that behavioral geneticists studied primarily adoptive families. Their major methodology compared the similarities and differences among adoptees, adoptive parents, and biological parents, and also between biological and adoptive siblings. Behavioral genetics, in addition, studied identical and fraternal twins, including those raised apart through adoption.[12]

While I had not known about behavioral genetics by name, I had heard of its more infamous practitioners. Francis Galton, the cousin of Charles Darwin, initiated twin studies in the nineteenth century where he discovered the impact of genes, but he also started the eugenics movement that advocated selective breeding and sterilization to "improve humanity." Gregor Mengele, a German scientist, used Galton's ideas to justify brutal experiments in Nazi concentration camps. In the 1960s, psychologist Arthur Jensen published controversial studies of *group* racial differences in IQ.

This history caused the field to fall into disrepute, but behavioral genetics in the last forty years has gained legitimacy as a science. It has rejected genetic determinism and focused on analyzing *individual* differences. Still, it was hard for me to forget this history and to look at behavioral genetics as a serious science. But their use of adoptive families as subjects led me to investigate further.

Behavioral genetics no longer sees genes and the environment, nature and nurture, as separate entities. Rather, people often select environments that fit their genetic makeup and the environment can cause genetic change, alterations that can be passed on to the next generation in a process called epigenesis.[13] In the past several decades, molecular genetics has begun trying to link specific gene sequences to individual behavior, but it has had only limited success. Nor has this science been able to link specific environmental characteristics with personal outcomes. Yet by using adoption and twin studies, behavioral genetics tries to establish correlations.

I found some of their general findings interesting:[14]

- Individual differences in behavior, personality, and cognitive abilities, including personality disorders, all have a genetic basis to some extent.

- Environmental influences—both inside and outside families— are at least as important in explaining individual differences.[15]

Rather than pursue generalizations about human behavior, however, I wanted to look at the empirical data gathered from adoptive families to see what it could tell me about my adoption experience.[16] Early in my search, one study caught my attention with findings that were unexpected and revealing.

Robert Plomin and colleagues at the Colorado Adoption Project studied 245 adoptees, placed in the first few months of life, and followed them for twenty years. Most of the subjects, in this and other behavioral genetics studies based on adoption before 2000, were white. They compared the children's cognitive abilities to those of their birth and adoptive parents. Cognitive abilities are measured by a number of tests—for verbal skills, reasoning, spatial visualization, speed of processing, and memory, among others. The researchers also compared the adoptive and birth parents to parents who raised their biological children (called a control group).[17] The birth parents' cognitive skills were tested before they placed their infants for adoption. Then the researchers examined the children, adoptive parents, and control parents a number of times as the children were growing up. At age sixteen, the adopted children and those in the control group were given the same set of thirteen tests of specific cognitive abilities that biological parents (birth parents and controls) had been given at the beginning of the study.

The findings were counterintuitive to the prevailing emphasis on the importance of nurture: The correlation between adoptive parents and their adopted children in cognitive ability was always near zero, even when they had lived together for sixteen years. For birth parents it was a different story. In early years the correlation in cognitive ability between birth mothers and the children they relinquished was only slightly higher than with the adoptive parents, but the correlation increased over time. By age sixteen, the adopted children were as similar to their birth parents whom they had never seen as were biological children (in the control group) who lived with their parents. When researchers looked only at verbal ability, adoptive children before age four had similarities to their adoptive parents, but this similarity completely disappeared by middle childhood. As they aged, the adopted children became more like their biological parents in verbal ability. The researchers concluded that environmental transmission from parent to offspring has little effect on later cognitive ability.

What did this study mean to me? It began to give me some insight into why the young boy whom I felt so close to and in tune with deviated from me as a teenager and beyond, making choices that were so different from mine. I had known that intelligence broadly defined had a genetic component, so I was pleased that he did well in elementary and middle school and did not have learning disabilities that would make his life more difficult.

Through the reunion, however, I learned that there was academic ability in Louise's family and that Marco's intelligence didn't just come from living with me. In Marco's picture necklace, Louise's brother Don, the squirrel-hunting social worker, is shown as a tall, jovial man in his early sixties wearing camouflage pants and purple socks. Before I met Don, Marco told me that he was the one most like me in his birth families. When I met him, Don told me that he had a Ph.D. in engineering. But Don had given up his employment as an engineer for an oil company. He felt called to do something for the poor and returned to the university to get a social work degree. He found a job working with Catholic Charities primarily in poor, black communities. Through genealogical research, Don discovered that his great-grandfather was a physics professor in France. These facts provided a probable genetic basis for Marco's cognitive abilities.

Marco was very verbal, which I had always presumed was due to my influence and that of my friends. Now, this study was telling me that Marco's verbal skills were more closely linked to the verbal ability of his birth mother and those in her family.

I discovered another study from the Colorado Adoption Project focused on reading performance which concluded that reading ability is influenced by both genes and environment.[18] Perhaps this study would help to explain why I loved to read and Marco never became a reader. The research looked at reading scores for adoptive children and their adoptive parents and how they changed over time. It compared them with those of a group of biological parents and children. The reading scores of both sets of parents were similar, but the reading performances of their children were different. In adoptive families, the reading abilities of parents and adopted children were not related. In contrast, in the biological families, the parent and child reading skills had a small but significant correlation. The authors concluded that parental reading had little impact on individual differences in children's reading performance. Individual differences in reading ability, they said, are more the result of genetic influence than of cultural learning.

I read a lot to Marco as a child, but not to influence his future reading ability. Rather, we both enjoyed the bedtime stories that I read to him every night even after he could read himself, and it strengthened the bond between us. I still have a trunk of children's books that Marco wanted me to save. I assumed that his bond with these books, his observation of all the books I had in the house, and my love of reading would translate into an adult who read. The behavioral genetics study gave me some insight into why this didn't happen. Finally, in his mid-thirties, as he is in the process of controlling some of his addictive behavior, Marco has started reading for pleasure. Perhaps nature (his genetic intelligence) and nurture (my love of reading) are beginning to merge.

I thought that I had an open mind about my adopted son's educational achievement. But like most parents, I had unacknowledged expectations. Because everyone in my family went to college, because I was a college teacher, and because I saw higher education as usually necessary for middle-class status, I just assumed Marco would be a college graduate. But by age thirty, he had less than one year of college education and we had learned that neither Louise nor Reggie had attended college. I discovered research which shed light on the educational congruence between Marco and his birth parents—research which indicates that personality factors may be more important than cognitive ability in educational achievement.

Using the Colorado adoption data, researchers looked at how parents in adoptive families and in control families rated their children at age four on what they call "an attention span persistence scale." This scale rated five kinds of behavior: The child (1) plays with a single toy for long periods of time, (2) persists at a task until successful, (3) goes from toy to toy quickly, (4) gives up easily when difficulties are encountered, or (5) gives up quite easily with a difficult toy. The first two criteria are correlated with academic achievement.[19]

I remembered that Marco as a child would go from toy to toy quickly and give up easily. In middle childhood, he was good at piano, art, theater, juggling, and sports, but he never persisted with any interest, dropping one and going on to the next. The researchers found that children at age four who had attention-span persistence increased their odds by 48.7 percent of completing college by age twenty-five. A lack of attention persistence (not the same as ADD, attention deficit disorder), a characteristic with a strong genetic component, did not mean that a person would not succeed in college, but worked against it.

If I had known about this research, would I have tried to intervene to improve Marco's concentration? But how could I have done this? At the time, there was little to guide me, but more recent research stresses the plasticity of the brain and how nurture can alter brain circuits.

Later (in the appendix), I look at the work of some behavioral geneticists who do research on the characteristics of birth parents and then counsel adoptive parents on how to intervene to maximize the expression of genes for positive behaviors and to minimize those for negative characteristics.

I found other behavioral genetic studies about what some psychologists now accept as the *big five* personality traits: extroversion, agreeableness, conscientiousness, emotional stability, and intellectual openness.[20] A Texas Adoption Project found that genetic correlations between birth parents and adopted children for these personality traits were smaller than for cognition, but similarities in personality traits also increased with age, even when birth parents and their adopted-away children had no contact. I found this study especially fascinating.

Researchers at the University of Texas began this study in the mid-1970s. They recruited families who had adopted through a Texas private home for unwed mothers.[21] All adoptions were in the first few weeks of life and were closed. The adoption agency had conducted routine personality tests on the birth mothers before they relinquished their children. The adoptive parents then agreed to take the same tests that the birth mothers had completed earlier. Both the adoptive parents and the birth mothers were primarily white and middle-class, but their personality profiles differed. Birth mothers had a much higher level of emotional instability and maladjustment, perhaps due to genetics, but more likely because most of the birth mothers were young and unmarried. Or a combination of genetic and social factors could explain this finding.

When the adopted children were age seventeen, the Texas researchers tested them, using the same personality tests that the birth mothers and adoptive parents had taken prior to placement. They also tested siblings who were biological offspring of the adoptive parents.[22] The researchers found "little if any personality resemblance between adoptive parents and their adopted children across a wide variety of personality tests."[23] But they did find a modest correlation between the adopted adolescents and their birth mothers, a correlation similar to that between the adoptive parents and their biological children.

This Texas adoption study also compared parental ratings of the personality characteristics of their adopted and biological children at seven years old and then ten years later. Although all the adopted teenagers seemed to be leading reasonably satisfactory lives, their parents' rating of them changed over time. At the younger age, parents rated their biological and adopted children equally favorable, but by the time the children reached age seventeen, they rated their adoptive children as less adjusted and less emotionally stable than their biological children. The authors speculate on three possible reasons for the change. A trait that is appealing in a seven-year-old (impulsiveness and extroversion) might not be so charming at seventeen. Or, some troublesome attributes may not emerge until adolescence. Or, adoptees may have a special sensitivity to their environment, reacting well to the warm, supportive atmosphere of the adoptive home in early childhood and not as well to the broader environment later on.[24]

Some thirty years after the original study, Texas researchers contacted adoptive parents and their adopted and biological children to look at life outcomes, especially educational and occupational achievement as well as personality characteristics. The adopted sons and daughters, now in their thirties, were on average doing well, but the biological children had done better. The biological offspring's personalities were more like those of their parents. There was almost no relationship in life achievements between unrelated adoptive siblings growing up in the same family, nor between adopted and biological siblings raised together. The researchers concluded that there is a moderate but enduring genetic influence in life outcomes for adopted children.[25] One must remember, however, that these studies measure only a limited number of characteristics and do not rule out the impact of nurture by adoptive parents.[26]

Thinking about these studies helped me understand some of my experiences. Marco, like many adopted and biological adolescents, rebelled in his teenage years. He never accused me of not being his mother and seldom got angry, but he began to live differently than I did. His life crossed not just race boundaries, but also class barriers. He was drawn to people who were poor and/or outsiders and who were not interested in school achievement. I would have approved of his expansiveness if it had not involved a steep decline in school performance, drug use, and, as I found out later, drug dealing as early as seventh grade.

At the time, I considered this as teenage rebellion and thought that the Marco I knew and loved would return. But he was gone for a long time. Some of his more positive traits, his warmth, sociability, and empathy, appeared from time to time in his late teens and early twenties, but too often he was self-absorbed and, more and more, he lied and hid himself from me. Only in his mid-thirties, as his drug and alcohol use has declined, has more of the wonderful Marco returned. I thought that Marco's intellectual ability, but also his musical, artistic, and theatrical skills, ones I value but do not possess, would lead him back to a mainstream life. They have not.

Perhaps all teenagers rebel in an attempt to figure out who they are independent of their parents. But with an adopted teenager, "who one is" may turn out to be closer to biological parents, even ones they have never met, with all kinds of positive and negative traits that are quite different from those of the adoptive parents. This research helped me see why it was hard for me, and for many other adoptive parents, to understand and guide our teenagers and young adults.

I'm not ready to abandon the theory of psychological loss through adoption as a partial explanation for the difficulties faced by adoptees, but it has to be supplemented by an understanding of the impact of genetics. Rather than focusing on the break between birth families and the children they place for adoption, as in the theory of psychological loss, behavioral genetics leads us to look at the continuities between birth families and adoptees even when there is no contact.

Yet, I wondered, how important were these personality differences between adoptive parents and their adopted children? After all, I'm quite different from my brother and sister, although we share half our genes and a family upbringing. My educational and occupational choices and my problems are quite distinct from theirs, and my life is completely different from that of my mother. Despite these dissimilarities, we all maintain a family bond. Most family members do, whether joined by biology, adoption, or as a blended family resulting from divorce and remarriage.

Many of us who decide to adopt know we can appreciate and love a child who is different from us. I liked that I could relate to and enjoy Marco's birth families despite cultural and social-class difference, and I found personal traits in them that I valued, such as the extroversion and cognitive ability in Louise's family and Reggie's artistic and musical talents, attributes they had passed on to my son. I concluded that differences in personality and cognitive traits had affected my relationship with my son, but had not damaged it permanently. Here, I have stressed the ways in which he disappointed and troubled me, but I always loved him. In chapter 6, I explore the close bond we formed in his youth and how that sustained us through the rough times. But another characteristic of Marco and his birth families did have a negative impact on our relationship—the predilection to addiction.

CHAPTER 3

Tarnished Jewels

Before returning to Louisiana for his second and longest stay, Marco had told me that Louise and Reggie seemed to be romantically interested in each other. He didn't like the idea, as he probably wanted Reggie for himself. Marco had always confided in me and continued to do so through phone reports. Upon his arrival, he found that Reggie had moved in with Louise, even though both were married to someone else.[1] Her husband, a Central American immigrant, often traveled to get construction work, so I wondered if he didn't know about this new arrangement. Whatever the situation, Marco told me that he found it hard to endure the ongoing turmoil. Louise, he reported, repeatedly kicked Reggie out, then took him back. Marco, who had always identified more with Reggie than with Louise, began to express disillusionment with his father: "Reggie has a lot of ambition, but he doesn't get anything done."

At about this time I was out of the country at a conference, so I didn't talk to Marco for three weeks. When I got home, I found out that Reggie seemed to be gone and Marco was working on a construction job with Louise's husband. Marco said he enjoyed living there now. "I love that man to death," he said of Reggie, "but I can't rely on him." Another time he said, "All that man likes to do is hang out and get high."

Marco wanted me to come for a visit immediately, saying that he wanted me to experience Mardi Gras. He also wanted the two of us to take a trip around the southern part of Louisiana, going to parks and bayous. "Everyone here just likes to stay put and watch television," he lamented. I was glad that Marco wanted to bridge his two worlds by appealing to our mutual love of travel, food, and new cultural experiences, but I didn't want to get caught in the complex dynamics of his Louisiana families. My decision was influenced, too, by my reading of memoirs and policy work about adoption, which presented reunion as something an adult adoptee did on her or his own. I'd already taken a more active role than any adoptive parents I'd read or heard about.[2] I said that I would come at a later date. If I had gone then, I don't think it would have changed the trajectory that followed, but I'll never know. I would now advise adoptive parents to be as involved in a reunion as their adult daughter or son requests.

For the next month, I got bits of information indicating that Marco's positive vision about reunion with his birth parents was fading. I learned he left Louise's house after a fight with her husband. The next time I heard from him, Marco was back in Reggie's hometown, where the two of them lived in the grandfather's old house with no water, heat, or electricity. Keeping warm with blankets borrowed from a relative, they were doing odd jobs around town for money and food and sometimes getting a meal from Reggie's sister, Dottie. Once Marco called me from Dottie's, where he had taken a shower and was doing his laundry. He asked me to send a copy of his birth certificate since he had lost his wallet with his California ID and his social security card.

I was distraught. I intuited that he was using with his father. His life in Louisiana had disintegrated into one similar to his life in Berkeley and Oakland—living on the edge, losing things, relying on the generosity of others. It was an addict's life.

Maybe I should have let Marco make his own way, but I decided that I wanted to reach out to family resources in Louisiana. I called Louise. She told me she had not heard from Marco and didn't know he was with his father. I decided not to mention her pledge to keep Marco away from Reggie. Even though I made no accusations and didn't blame her, she became defensive: "I didn't know what I was in for when I invited Marco to live with me. I didn't know he had anxiety about applying for a job or that he would refuse to cut his dreads, something my relatives required to give him a job."

I knew that Marco had problems with anxiety, but I had attributed this to his substance abuse. This assumption was later affirmed by a therapist who called Marco's anxiety "garden variety," likely a result of his addiction. But Marco could also have been influenced by my anxiety. I now felt guilty for facilitating his return to Louisiana, and this guilt justified further intervention. I asked Louise for the phone number of her brother Don, the Catholic social worker. I hoped that he could help Marco.

Don was open and friendly on the phone. He told me how impressed he was with Marco when he met him—how smart, articulate, and full of life he seemed. Don hadn't known about Marco's addiction and was concerned when I told him where Marco now lived. "You can buy any kind of drug in that town," he said, his knowledge gained from having grown up and worked there. Don promised he would look into locating a rehab program and would talk to Marco. I didn't know if he could make a difference, but I gave him the phone number for Dottie, Reggie's half-sister, as a link to Marco. I was glad that I had called him.

But Don's intervention proved unsuccessful. He found a free rehab program and called Marco, but Dottie later told me that Marco was furious with me for telling Don about his addiction. Marco never said anything to me, which reinforced my belief that his addiction was serious. I didn't hear from Don again, but Dottie became my ally and we regularly talked on the phone.

In Marco's picture necklace, Dottie is a short dark woman with a prominent gold tooth, half-hidden behind Reggie, her taller, lighter-skinned, and more attractive brother. But the woman I encountered on the phone had a powerful presence. She seemed to fit my stereotype of a strong black woman, one who was comfortable talking to me despite the distance between our worlds. I had expected that I would get along better with Don, a professional white relative, but I was relating more easily with this black woman who had little formal education. I felt linked to Dottie as a mother and because of her knowledge about addiction.

Because she was a church lady, with gospel music on her answering machine, I had conjured up an image of her living in a rundown Victorian house, perhaps bigger than my small, three-story house in Berkeley. She told me that sometimes her adult son came home for a while. I'd assumed Marco stayed in this son's room or in a guest room.

Later, on my trip to Louisiana with Marco, we visited Dottie. We drove up to an exceedingly small, one-story house with a pickup truck in front, smaller than Louise's trailer or Reggie and Jeanette's townhouse. Dottie met us at the door, now seeming shy. We entered a small living/dining room shut off from the rest of the house by a curtain. I presumed that a kitchen, bedroom, and bathroom lay behind, but no one went back there during our visit, so I never found out. When I asked Marco where he had slept, he pointed to a narrow space between the round dining table and the couch, where only a small single mattress or a blanket roll would fit. In this cramped space, two furnishings stood out: a preserved round oak table with a set of carved oak chairs, more beautiful than the ones I own, and a large wall hanging, woven in shades of brown, tan, and white, which depicted the heads of young black girls, some holding their hands in prayer. After I complimented her on these furnishings, Dottie warmed up. But as she talked I experienced more differences in our backgrounds.

Dottie was only in her forties, but she had not worked for some time and lived on disability payments. She had just returned from her first-ever plane ride to visit a relative in Texas and recounted her terror that turned into relief and amazement. Marco had told me that Dottie was an evangelical Protestant, who attended services in a community center outside of town where worshipers often spoke in tongues. Every week, Dottie told me, she burst into tears, so grateful that this church had saved her from the ravages of addiction.

Taken aback by the distance between Dottie's life and the middle-class lady I had imagined, I was again chastened by my own class and race assumptions. I realized I was receiving an important education which would deepen and complicate my understanding of social class and of Marco's biological heritage.

Marco got to know Dottie long before I met her. She, too, wanted to save Marco. When he told her that he used to have a problem with drugs and alcohol, but that he no longer did, she hadn't believed him. She invited Marco to stay in her house, fed him, and bought him new clothes, which he needed. She took him to church, set a curfew of 10:00 p.m. on week nights and midnight on the weekend, and forbade any drugs or alcohol in her house. I was grateful but asked myself, "How long will this last?"

A few weeks later, Dottie called to tell me that she had evicted Marco. She couldn't stand him using her house as a pit stop. Every few days, she said, he stayed out all night, coming home late the next day for dinner and a shower. She knew that he was using drugs and hanging out with other users. This news fueled my despair, but I thought she did the right thing. She continued to keep me up-to-date on Marco's situation, telling me she saw Marco once looking ratty and doing odd jobs that paid only two to three dollars an hour. One day, she said, "I was out with a friend and I saw him sweeping a parking lot with a small broom, like some Asian coolie. I was embarrassed, since many in town know he's my nephew."

I presumed that Marco was back living with Reggie in the shack.

I now knew that this reunion with his Louisiana families was not saving Marco. Rather, his desire to relate to his father seemed to be sending him deeper into addiction. I was glad that this aunt, who could identify with what it meant to be black and who knew more about addiction than I did, was trying to help. I now felt less anxious and less responsible, but sadder and more resigned.

The distance between Marco's world and my own in middle-class Berkeley was placed in stark contrast a few weeks later on a Saturday night when I was hosting a dinner party for eight people to introduce friends visiting from Scotland. Cooking and entertaining were two of the few interests I shared with my mother, whose skills I had emulated. I also liked how my mother furnished our home. In my living room I use her lamps, a large mirror, and a wall hanging. This evening I was using one of her tablecloths, her china, and her silverware, As a single, working woman, however, I didn't entertain often, for it took a lot of energy and concentration to set the table, do the cooking, serve the guests, and help keep the conversation going.

As my guests were gathering for wine and appetizers in the living room, and I was introducing people to each other, the phone rang. It was Marco calling from a Louisiana jail.

"Mom," Marco wailed, "you have to arrange for me to call you collect."

"I can't do it until tomorrow," I replied, and hung up the phone. I had stopped bailing Marco out of California jails for small infractions, but this was a different state and I needed to find out more about the charges. The next phone call was from Dottie. I told her that I wouldn't bail Marco out. "Good," she said. "This small town jail isn't a dangerous place." I was relieved. Louise also phoned that evening and seemed surprised when I told her I wouldn't bail out Marco.

I said nothing to my guests about what was happening. Even if they had been closer friends, I would not have dampened the lively party with my troubles. As I write this, I ask myself why I hadn't turned off the phone so that I could concentrate on my guests. I realize that despite my anger with and discouragement about Marco, I was deeply involved with him. He was my son and I couldn't ignore him.

The next day, Dottie called to tell me that Marco's arrest was for shoplifting, a crime for which he had previously never been arrested. According to Marco, he and Reggie had taken sporting goods from Wal-Mart, selling them around town for half-price, presenting this as a service to the poor. "I could be in jail for a year and a half," Marco said as he pleaded with me to bail him out. Since I knew that Louisiana had a higher percentage of incarcerated people than any other state, I believed him. I told him that I needed to think about it. Louise soon called and said that she wanted to bail Marco out, but that $170 was a lot of money for her.

I said, "I will help with the bail, but only if you drive him directly to the rehab center that Don found." She agreed, but I heard nothing for several days. She didn't answer her phone, and her message machine was turned off. In frustration, I called Dottie again. She told me Louise had posted bail for Marco and he was no longer in town. Where had she taken him? Why had she not asked me to send the money?

When I called Louise again, Marco answered. He had work painting houses and had paid the bail money back to Louise. I hoped this was true, for it would mark a positive change. Louise obviously had not taken him to rehab. "Marco, are you doing anything about your drug and alcohol problem?" I asked.

"What problem?" he replied. "I don't like to talk to you since that's all you want to talk about."

"I hope that your addiction has not turned off your family there the way it did me," I said, angry and frustrated. He changed the subject.

It was Mother's Day. For the first time in twenty-six years, Marco wasn't spending the day with me, but with his birth mother three thousand miles away. Knowing that I might feel lonely, I planned a busy day starting with brunch with my friend and surrogate daughter Julie, who is ten years older than Marco, and whom I've known since she was seven. She'd lived with me for several years after Marco moved out. After lunch, we walked in the botanical garden, overflowing with rhododendrons, azaleas, and other spring blooms I love.

But all I could think about were the Mother's Days I'd spent with Marco. He rarely had money to buy me a present or take me out, so his gift was to spend the day with me, doing what I wanted. Yet, I enjoyed myself only when he was engaged too. We both got pleasure from good food, but restaurants were crowded on Mother's Day, and as he advanced into his twenties, I resented having to pay. Sometimes he cooked breakfast for me or barbequed for dinner, but I'd have to stock the ingredients. We both liked to hike or take the dog to the beach, but since he had lost his license, I'd have to drive. I was having a hard time accepting that he wasn't turning into the responsible adult I had expected he would become.

Then, several years earlier, I had hit on the perfect activity—going by public transportation to visit a San Francisco art museum. He liked going to museums with me, and I could enjoy being with Marco without having to drive or pay, since I was already a museum member. But last year he had slouched behind me as we entered the San Francisco Museum of Modern Art. I was an older white woman dressed conservatively accompanied by a young black man with long dreadlocks in baggy jeans and a white tee. Did he hesitate because he felt as conspicuous as I did?

I had started taking Marco to view African American artists when he was eight or nine. Although I came to love some of these artists, whom previously I had known little about, especially Jacob Lawrence, Romare Bearden, and Aaron Douglas, Marco often preferred other artists in the museum. On Mother's Day that year, I wanted to see an exhibit by Robert Bechtle, a Bay Area realist painter. I warmed to Bechtle's paintings because they often featured familiar locales and evoked the starkness and isolation of American life in a subtler way than did Edward Hopper's work. Marco became enthralled. He peered closely at the paintings.

"Look at the subtle shading on the houses and autos," he pointed out. "These paintings make me remember my own feelings, happy and sad, about such cars and those neighborhoods, something I would never get from a photograph."

He was so intrigued that he forgot his unease. Perhaps his interest indicated a male love of cars, but I took it as more than that—an appreciation of art. For a few hours, our interests and worlds merged with each other. This was the son I had so enjoyed when he was a child, the son I longed to have as an adult.

But this Mother's Day, I never heard from Marco. The Monday after, I called Louise at work. Before I could say anything, she said: "Marco didn't call you yesterday because I no longer have long-distance service for outgoing calls. Please call him; he feels bad." I called and he answered, "Happy Mother's Day." I felt better, but wondered whether this technicality was the real reason he hadn't called. My doubts were reinforced when he called me from Louise's house two weeks later on Memorial Day weekend. Louise was by his side and I noted to myself that neither of them mentioned anything about the phone service being restored. This was the sort of detective work I'd fallen into over the years, trying to figure out what to believe or not, a process familiar to those with addicts in their families.

Marco started with chitchat and questioned me about my life, but I could tell by the tone of his voice that something else was coming. "I know you don't want to send me money," he plunged in, "but the household is in crisis." He told me Louise had a shock that morning when her ATM card was rejected. The bank told her that someone had written a $1,000 check on the $200 she had left in the account. The bank would replace her money once they'd investigated but they couldn't do it until Tuesday. Ed, Louise's husband, hadn't been home and didn't answer his cell phone. Louise suspected he had written the bad check. "Put Louise on," I said.

"I would never have called you to ask for a loan," Louise said, "and I know nothing about wiring money, but Marco insisted. My whole family is in Florida, and I have no one I can ask to borrow some cash. I hate asking you for money, but we'll repay you next week."

I quickly went over in my mind factors to consider before making a response. Again I had to become a detective. Louise said that she would keep Marco away from drugs and alcohol, but she hadn't. She told me that she would send him to rehab when she bailed him out of jail, but she hadn't. There was a discrepancy between what she told me and what her brother, Don, reported. Louise said that Marco had left her house after the first six weeks of living there, because he had a fight with her husband. Don told me that Louise had asked Marco to leave because she couldn't be around someone who used drugs. All this undermined the positive impressions I'd had of Louise, but I reminded myself that addiction was a disease, not a moral failing. She seemed so desperate, and I'd feel guilty if I refused her plea when I'm so much better off.

"Would $200 be enough?" I asked.

"More than enough," she replied.

I wired the money, but I was still conflicted and suspicious. Marco called twice in the next week. "I'm trying to get the money together to pay you back," he reported, without my asking. I didn't remind him that I had sent the money to Louise, not to him. Why did he say nothing about her? Once, when I called her at work to give her a message for Marco, she said nothing about the money. A question crossed my mind: Had she been willing to front for him to get him money for drugs and alcohol? I rejected the idea, for I didn't think a mother would do this.

A week later, I told the story in my Failure to Launch group. For more than a year, I'd been going to this group organized by a therapist for parents whose adult children were not progressing toward independent lives. Almost all of us were parents of sons who had addiction or mental health problems or both. Because many of these parents had more experience with drugs than I did, and because many of them had faced greater problems with their biological sons than I had with my adopted one, I'd learned as much from them as from the therapist. Once when I reflected that parenting might have been easier for me if I had not adopted, a couple replied that they often wished they had adopted since there was so much mental illness in both their families. That evening, in response to my story, the conversation went like this:

"It was worth $200 to find out what is really going on," one man said.

"You must see something I don't," I responded. "What is going on?"

"I think everyone in the household is using," April, the therapist, interjected.

"How could a mother collaborate in her son's addiction?" I asked in shock.

"She could if she was an addict too," April replied.

A week later, after Louise again had ejected Marco from her home, he confirmed this. Even the therapist was taken aback when I reported at the next meeting that according to Marco, Louise spent hundreds of dollars a month on cocaine. She drove Marco to the "bad" (black) parts of town where a white woman by herself couldn't go. She paid, he scored, they used together.

Learning that both Marco's birth parents abused drugs saddened me. After he returned to California, Marco told me he was not surprised to learn of their addictions. "I always felt something was different about my brain," he said, implying that his tendency to addiction was inherited. I later learned that Marco's assumption could well have been correct. Pathways in the brain mediate between genes and behavior.[3]

I had previously dipped into the vast literature on addiction. I learned a great deal from Dr. Robert L. DuPont, who, in *The Selfish Brain*, defines addiction as a complex, lifelong biopsychosocial brain disease.[4] I also liked *Clean: Overcoming Addiction,* by journalist David Sheff, a book based on extensive research into the scientific literature. Sheff defines an addict as anyone who continues to use drugs or alcohol in spite of harmful consequences.[5] Like DuPont, Sheff views addiction as a brain disease. He discusses research showing that addicts' brains are different *before* they use drugs, not just afterward. There is evidence, too, that addicts' brains are different from birth.[6] But the use of drugs and alcohol further alters the brain. Sheff asserts that the disease of addiction—like many diseases—has a genetic component that is inherited.[7] It's the predisposition, not the disease itself, which is inherited, and not by every child and not in every generation.

Inspecting Sheff's footnotes, I discovered that he based many of his conclusions on behavioral genetics research on addiction in twins. Using twin studies, behavioral geneticists have concluded that genetic factors are involved in vulnerability to drugs and alcohol use, initiation of substance use, continued use, and addiction. For men, genetic factors play a larger role in persistent substance use, while in women, genetics is more important in initiation of drug and alcohol use.[8] I then looked more closely at other behavioral genetics studies based on adoptees.

A recent investigation by behavioral genetics researchers at the University of Minnesota compared parents' drinking to that of their offspring in a longitudinal adoption study.[9] Using a large sample of 409 adoptive and 209 nonadoptive families,[10] they compared the drinking behavior of parents and children for three time periods, starting when the youngest child was eleven and ending when the oldest was twenty-eight. Unlike other studies, this one did not have data on the biological parents of the adoptees. The researchers reported a moderate parent-child correlation in drinking, with a higher correlation in the nonadoptive families. The most interesting finding, however, was that the correlation between parents drinking and that of adopted sons and daughters *decreased* over time, especially as adopted children reached their mid-twenties, while the correlation *increased* in the biological families. The study concludes that as young adults leave home, the impact of the home environment decreases and the influence of genetic factors increases in both adoptive and nonadoptive families. It was interesting to me that the findings of this study paralleled those of research I had looked at earlier, on the declining influence of adoptive parents on the cognitive and personality characteristics of their adopted children as they grow up.

Another study on the impact of genes on alcoholism used data from adult males adopted at birth in the early 1990s from various agencies in Iowa.[11] The researchers compared male adoptees whose birth parents were known alcoholics with males from the same adoption agencies whose birth parents did not abuse alcohol. Male adoptees from an alcoholic biological background were almost twice as likely to become drug abusers as those whose biological parents did not abuse alcohol.[12] This study, however, confirmed that genetic predisposition is not the only factor leading to substance abuse, since almost one-third of adoptees whose biological parents did not abuse alcohol became addicts. The study demonstrated that there are a number of other pathways to substance abuse in adoptees—through the impact of alcohol and drug use by adoptive parents, through conflicts and instability in the adopted home, and indirectly through a genetic risk for personality disorders that can facilitate substance abuse.[13]

An additional conclusion of this study was that a similar group of genetically transmitted factors is involved in both alcohol and drug abuse, a finding validated in more recent studies. The journal *Behavior Genetics* published a study in 2006 (one not based on adoption data) concluding that adolescents' genetic vulnerability for substance abuse operates in a general way across a variety of substances.[14] Marco had told me that his birth parents were addicted to crack cocaine, but that he used a variety of drugs and alcohol. I believed this until I read the afterword he wrote for this book. Still, I found this study interesting, and it probably has relevance to others.

In support of the idea that genetic predispositions are not necessarily substance specific, neuroscientists have identified common pathways in the brain for the rewards and addictive actions of all drugs of abuse.[15] Neuroscience research has indicated that the neurotransmitter dopamine in the brain may be associated with the degree of pleasure that is experienced on first use of a drug, which in turn could be related to risk for further use and potential addiction. Differences in dopamine production are governed both by environmental influences and by genetic variation.[16]

A national Swedish adoption study, based on large numbers, included adoptees in Marco's situation of having both biological parents with addictions.[17] The Swedish study concluded that 4.5 percent of adopted individuals abused drugs compared with 2.9 percent in all of Sweden for the same years. Among adopted children the risk of drug abuse for those who had no parents with substance abuse was 4.2 percent; for those who had one drug-abusing parent, 8.2 percent; and for those where two biological parents abused drugs, 11.9 percent of adoptees had a risk of drug abuse. That is, having two biological parents who abused drugs tripled the likelihood that an adopted child would also do so. The study further found that drug-abuse risk was higher for male adoptees than for females.[18] I could see, therefore, how Marco's biological background put him at risk for substance abuse, although the risk factor was still quite small.

Going back to the Iowa study, I learned of a second genetic pathway to drug abuse through the inheritance of personality factors: an antisocial personality in a biological parent can lead to a similar syndrome in the adoptee, and such a personality factor can predispose one to addiction. Could Marco with his extroverted personality and good social skills be considered antisocial? I discovered that this personality constellation has a very specific meaning. In the 2013 Diagnostic and Statistical Manual of Mental Disorders (DSM-5), anyone over age fifteen who met at least three out of seven criteria could be classified with an antisocial personality disorder. Marco met four. Here are the seven:

1. Failure to conform to social norms with respect to lawful behavior as indicated by repeatedly performing acts that are grounds for arrest (yes for Marco);

2. Deception, as indicated by repeatedly lying, use of aliases, or conning others for personal profit or pleasure (yes for Marco);

3. Impulsivity or failure to plan ahead (strong yes for Marco);

4. Consistent irresponsibility, as indicated by repeated failure to sustain consistent work behavior or honor financial obligations (strong yes for Marco);

5. Irritability and aggressiveness, as indicated by repeated physical fights or assaults (no for Marco);

6. Reckless disregard for safety of self or others (no for Marco);

7. Lack of remorse, as indicated by being indifferent to or rationalizing having hurt, mistreated, or stolen from another (no for Marco).

I discovered two studies that separated the criteria for antisocial personality disorder into two components—one which the research labeled delinquency and the second which they labeled aggressiveness. Delinquency included the first four criteria, which fit Marco, while aggressiveness based on criteria 5 and 6 did not. Criterion 7 was said to be hard to measure and was eliminated. The researchers concluded that both delinquency and aggressiveness predict risk for cannabis, cocaine, and alcohol dependence, but the genetic pathways may be different.[19]

Indicators of an aggressive personality can be identified in early childhood, and the genetic correlation remains constant as the child grows up. Delinquency, in contrast, begins only in early adolescence and the genetic component increases until age sixteen.[20] Delinquent teenagers with antisocial personality traits did not evolve into healthy adults. Rather, they continued into early adulthood to commit low-level crimes, to have substance abuse and mental health problems, and to continue impulsive behavior. This pattern of delinquency fit Marco's behavior.

Marco's delinquent, but mainly nonviolent, behavior, starting in adolescence and continuing into adulthood, seemed to mimic that of Reggie, his birth father, as did his addiction. Perhaps Louise, the birth mother, also shared their impulsive failure to plan ahead, for another researcher reported that impulsivity, without delinquency, has a genetic basis and is linked to addiction. This correlation is higher for those who abuse substances than for those who just use.[21] Again, *correlation* does not necessarily mean *cause*. Delinquent behavior could lead to substance abuse but the reverse could be true as well.

Discovering that Marco's birth parents abused drugs, and learning that Marco's substance abuse most likely had some genetic basis, relieved me of some guilt. However, learning that genetics is only a precondition for addiction, one that can explain only 50 percent or less of substance abuse, and one that needs to combine with other environmental factors to create an addict, made me reconsider my own responsibility. I needed to do more research on the impact of the adoptive family and the environment on addiction. Moreover, behavioral genetics gave me little insight into how to deal with Marco in the present. Right then, I had to respond to a new crisis in Marco's conflicted relationships with his birth parents.

One Friday afternoon in early June, when Marco was still living with Louise, she dropped him off to see a supposedly-now-clean Reggie, living again with his wife and some of her relatives. Louise said she would be back in a few hours to pick up Marco, but she never returned and was unreachable by phone all weekend. Jeanette and Reggie let Marco sleep on the couch in the crowded house. On Monday, Jeanette reached Louise at work and she agreed to speak to Marco. Over the phone, Louise blamed him for her relapse into cocaine use after years of sobriety; she said that she wanted him out of her life. I was surprised by her action since I had been impressed with Louise's parenting. Her other children were leading more productive lives than Marco and seemed to have no problems with addiction. They may have escaped the genetic predisposition, but Louise as a single mother, with fewer economic and educational resources than I had, although with more extended family support, had raised two healthy kids. I repressed a despairing hunch that Marco might have been better off if she had kept him.

Later, Louise called Marco to apologize, saying she had been treating him more like a friend than as a son. But he told me "the damage has been done." Would I accede to Marco's request to send him a bus ticket so that he could come home?

I know that adoptees often feel that their birth mother has rejected them. Now here was a rejection not by a teenage birth mother, but by a mature adult. I felt unbearably sad for him. How could I say no when his birth parents had disappointed him?

But he had made bad choices over and over again and he was an addict, so I hesitated. Despite all the phone calls, I'd had a carefree, happy spring, and I wasn't looking forward to his return. I wanted to uphold the pact I had made with myself and had pledged to the therapy group: never agree to a request for money from Marco without at least saying I had to think about it. I told him to call me the next morning.

I wrestled with myself all evening, using several friends as sounding boards. Then I drafted an e-mail to April, the therapist who led the Failure to Launch group. Writing to her clarified my thoughts. I had decided that when Marco called, I'd say I was not willing to buy a ticket unless he went into rehab in Louisiana for at least a month and preferably longer.

Then I would send him a ticket to come home. Or if he earned a ticket himself, I'd feel better about his coming back. I would say that I was sorry about his disappointment in his birth parents, but he had to start taking some responsibility for his life. Marco called early in the morning and I stuck to my resolve. He swore at me and banged down the phone. I was consoled by April's supportive e-mail which soon followed.

Forty-eight hours later, I got a cheerful phone call from Marco. "I'll be at the Oakland bus station in two hours; can you pick me up?" he asked. Louise had bought the ticket, so he had not met my condition, but I was happy that he was coming back home. I saw that I could not escape the roller-coaster ride of my feelings about him—from my strong bond with him to my desire for less responsibility, from joy to pain, from hope to despair. These emotional swings in my feelings had plagued my relationship with him for the last ten years.

I fetched him from the bus, gave him some dinner, and made it clear he could not stay with me. "No problem," he replied. His Jamaican friend was expecting him. I had vowed to myself that I would keep dinner conversation light. I wouldn't get openly upset with him or be critical. He presented his Louisiana stay, along with the bus trip back, as a young man's grand adventure, the shoplifting as a Robin Hood gesture. Were these his real feelings or a rationalization to hide his disappointment in his father? Either way, I found his remarks hard to take. But I said nothing.

My fear that Marco would limp back to California a worse addict than when he left had proved to be true. Marco's addiction landed him back on my doorstep. The picture necklace lay lost and forgotten, as those in the photos turned from shining stars to tarnished jewels—relatives with major challenges.

Marco never indicated any regret at having his Louisiana families in his life. He kept in touch with Reggie, his wife, and one of Reggie's daughters, his half-sister, and a few years later he returned to Louisiana for a shorter stay, the one where I visited him. But in ensuing years, Marco let these relationships lapse. Sometimes, his birth mother or his birth father's wife called me to try to connect with him. I'd have liked more contact, with the birth families. Having more contact, I thought, would deepen my connection with Marco, permitting me to better understand and relate to his traits which were different from mine. But I found it hard to maintain these relationships when Marco let them slip away. Still, I found comfort, and maybe Marco did too, in the knowledge that these family ties could be revitalized.

For now, I was Marco's sole mother again. In terms of an emotional bond, nurture had won out over nature, as it does for most adoptive parents. Adult adopted children usually renew their bonds with their adoptive parents after a reunion with birth parents.[22] But I no longer believed that nurture was determinative in adoption, or that genetic and prenatal heritage could be ignored. I now had to look at how our household and the environment in Berkeley had interacted with his genetic predisposition to addiction and his inherited personality to produce the adult he had become.

CHAPTER 4

Failure of an Alternative Family

I wish I could say that I provided such a wonderful environment for Marco that it offset his genetic predisposition for addiction. In fact, I was unable to protect Marco from trauma in his early years. The hostile breakup of my communal household when Marco was three and a half was as upsetting to me and to him as a divorce in a nuclear family.

I moved to Berkeley in 1974, seven years before I adopted Marco. Berkeley is a small, dense city with about 110,000 residents,[1] a city of hills and flats across the bay from San Francisco and tucked next to the larger city of Oakland. Berkeley has a relatively stable white population of about 60 percent, a declining black one (19 percent in 1990, 15 percent in 2000, and 10 percent in 2010), and an increasing percentage of Latino and Asian residents (11 and 19 percent respectively in 2010).[2] The small African American population of Berkeley is bolstered on its southern border by the larger and more multiracial city of Oakland, which holds four times more residents, about 30 percent of them black. Both cities have a small but growing number of people who officially identify as mixed-race—6 percent in Berkeley and 4 percent in Oakland. This racial and class diversity was in the background, something I liked but did not focus on.

I chose to live here, even though I always worked elsewhere, because of its intellectual and political character. "Berkeley" signifies one of the most progressive cities in the U.S., and it is known as a center of academic and scientific achievement. Sixty-eight percent of residents over age twenty-five have a bachelor's degree or higher, an extraordinary number obviously related to the presence of a famous university.[3] But that leaves one-third of the residents without a college degree and creates a city with the widest gap between rich and poor of any city in the Bay Area, more than San Francisco or Oakland.[4]

Berkeley also has a counterculture, not just one that is politically left, but one that is known for its alternative lifestyles and the free use of drugs. I was never part of the drug subculture, and it had little impact on my world, but I was attracted to Berkeley's heritage of left activism and experimentation with alternative household and family forms. I lived for short periods in two communal households before I formed the one into which I brought Marco.

I loved, too, the combination of an urban and university environment in a setting of natural beauty. My Berkeley epitomized who I was and who I wanted to be.

I formed my communal household in 1977 with Marcia and James, she a public health doctor and he a social worker, who had recently moved to the Bay Area. I had met Marcia through a mutual friend and we soon divulged that we were both interested in alternative living. I explicitly said to them and others that I expected to become a parent. At age thirty-seven, I wanted to settle down and have a child, but I didn't have a partner and I held negative stereotypes of single mothers. Ideas then current about communal living appealed to me, especially since I thought that bringing up a child in a communal household would avoid what I perceived as the problems of single motherhood. I envisioned a return to a more extended family form, but one based on friendship, not biology or marriage.

Both the feminism and the new left[5] with which I identified sought to create work collectives and communal households, usually separate in urban areas. Although reports differ on how many communal households existed in Berkeley in the 1970s, there were many, and households like ours were common.[6]

Marcia and James were a stable unmarried couple who had no extended family on the West Coast. James wanted children, but Marcia was ambivalent. He hoped, I think, that my clear commitment to mothering might rub off on her. It did. Since we had the same ideological and political positions on social issues, I assumed that we had the same expectations about sharing a family life.

We bought a large brown-shingle duplex as tenants-in-common; they renovated and occupied two floors upstairs and I had the downstairs flat. Real estate was relatively cheap in the 1970s; I borrowed part of my down payment from my parents, and James subsidized the rest. I worked off my debt to him by paying more than my share of the monthly mortgage payments.

We each had our own entrance with an inside stairway connecting us, permitting us to visit back and forth. Our major commitment was to share dinner four or five nights a week, each of us individually taking turns cooking. This way of living satisfied my desire for domesticity and eased my fears about either sinking into becoming a housewife in a nuclear family or ending up as an isolated single woman.

Finances were kept separate, and there was no sexual sharing. Since we came out of the politics of the left and of feminism, rather than the counterculture, none of us used drugs or drank beyond a glass of wine or beer with dinner. Did any of our friends ever smoke a joint at our house? Sure, but not regularly. For a commune in the 1970s, we were conservative. Like most communal households of the time, ours was white, middle-class, and college-educated.[7]

For the next four (childless) years, I flourished here, forging a life that suited me as a single woman. I lived with friends, but I had privacy and independence. Our household fostered a network between feminists and others who wanted to keep alive the culture and politics of the sixties. As I worked through the obstacles to childbearing and adoption during these years, I thought only abstractly about family in the future.

Our success at building a community sustained the household during these childless years. Our home hosted political meetings, study groups, consciousness-raising sessions, and book talks on feminist and progressive issues. We also hosted holiday gatherings and provided a place where out-of-town friends and acquaintances, often intellectuals and activists, could stay for a few days when they visited the Bay Area. When a woman friend—who had long been active on the left—bought a house next door, we became even more of a political center.

According to sociological criteria, ours was a genuine community, for we were at the center of a web of affect-laden relationships which crisscrossed and reinforced one another and we shared a culture with a set of common (left and feminist) values, history, and identity.[8]

Our household lasted almost eight years, much longer than the average two to three years for residential communes in the 1970s.[9] The short life of other communes probably reflected the youth of the residents and the difficulties of communal living in a culture that is so individualistic. The longevity of our communal household was rooted in our settled lifestyle—we all had jobs and were embedded in a stable community. But success at creating community did not translate automatically into an ability to form an alternative family.

My expectation was that after we had children we would become more family-like, but we never talked about our concrete goals, and probably theirs were as vague as mine. This belief was reinforced after Marco arrived. Marcia immediately got pregnant, and she delivered a baby boy nine months later. I envisioned three adults now collectively caring for two children. I'm not sure why I believed that three ambitious people in their late thirties and early forties, raised in nuclear families in an individualistic and competitive culture, were going to accomplish a different form of child rearing.[10] Hard as it is for a couple to meld parenting styles and expectations, it is even more difficult for friends to do so. But it seemed as if many people in Berkeley in the 1960s and 1970s were talking about extending collectivity to raising children. But few put the ideology into successful practice, probably for some of the same reasons that my attempt at a communal family failed.

Contrary to my ideology, for the first nine months after I adopted Marco, I didn't want to share the baby. Because of an academic fellowship, I could stay home with few work responsibilities and no fixed hours. Marco was an easy baby and I could pay for occasional childcare and babysitting. I didn't ask much from my housemates. Mostly, I valued the living arrangement for the built-in company and sharing of meals. Still, I expected that once their baby was born, we would share childcare. Billy's birth coincided with my return to full-time teaching, which included a long commute. Marco was in day care while I worked, but now that I was under more pressure, I'd hoped for more help at home. Just as we took turns cooking, I imagined that we would do the same with care for the children.

Perhaps I should have taken it as a warning sign that Marcia and James married before their son Billy arrived. They each had health insurance from their jobs, so that was not the reason for their decision. I might have questioned what it said about their commitment to an alternative family. Later, I looked on with dismay as they morphed into coparents willing to share only minimal childcare with me, their single parent friend who lived downstairs. But I hadn't wanted to share Marco when he was an infant, so how could I complain? Later, I realized we'd never discussed our beliefs about child rearing. We'd never talked about our expectations of each other or about how adding children might change the household dynamic. I actually had read little about parenting, probably for the same reason that I didn't read about adoption. Reading belonged to my work and not to my personal life.

I didn't consider that including an adopted, biracial child might make alternative family formation more difficult. We didn't talk about this or what it might mean to raise two boys with only one father. My father had been nurturing to all his three children, more caring and psychologically in tune than my mother, so unconsciously I thought James would be so too.

I sensed that I was more committed to the household than Marcia and James were, and I was afraid to rock the boat. Had I been more assertive, perhaps we could have worked out some of our differences or at least avoided the hostile ending. But maybe not.

Because I felt needy and one down to their coupled united front, I established a pattern of avoidance when I disagreed with one or both of them. Even before Billy was born, I began to resent James's dogmatic advice about how one should raise kids. For example, he believed that when a child woke in the night, you should let her or him cry. I disagreed, but rather than tell James that I didn't want his advice, I reverted to a childlike role. I'd sneak around the powerful *parent* to do what I wanted, feeling ever more resentful. I put a baby monitor in Marco's room, and at the first peep, I'd wake and go to him, hoping that the noise hadn't penetrated upstairs. This conflict avoidance had been established in my relationship with my mother and continued with male partners. It hadn't worked there, and it turned out to be ineffective in this relationship too. But this is me talking now; at the time I wasn't aware of these interpersonal dynamics.

Another problem was that I hadn't considered what it would mean for Marco to live in a household with a white man who was not his father. I hadn't expected that James would be more important to my son than Marcia. She had greater empathy for Marco and interacted with him more, but Marco clearly wanted attention from James.

When he was around three years old, Marco said to me, "I don't have a father, do I?"

"No," I replied.

"But I want one," said Marco.

"Some kids have fathers," I said, "but many don't."

I named several children of single mothers, but Marco responded, "Billy has a father."

This was the first time it dawned on me that our household might not be working for Marco. I saw that he might feel the loss of a father more than if he had been living only with a single mother. One Saturday afternoon, James came home when I was giving both boys, now ages two and three, a bath together. They were playing nicely until James rushed in, picked up Billy, and said, "Let's go upstairs and play."

Marco started splashing and hitting Billy, receiving an immediate reprimand from James, who seemed insensitive to Marco's feelings. Nor did James give any consideration to me. I had taken care of both boys for more than three hours and could have used a break.

I brought it up in a joint therapy session, and James apologized. But as Marco would say as an adult, about his relationship with Louise, the damage had been done.

One weekend, when Marco was three, a black African man, who had been introduced to me by my father, was visiting Berkeley for a conference and came over for lunch. Afterward, Marco asked shyly, "Was that my father?" I was startled.

"No. Your birth father is African American; he lives in Louisiana, not in Africa."

We had a picture of Marco's white birth mother, and I must have told him that his birth father was black. Or had Marco deduced that to get to his own light brown color, he had to have a black father? Perhaps his question indicated how few times we had a black man in our home. Certainly it revealed Marco's continuing curiosity, and perhaps a sense of loss, about not having a black or male parent. Sociologist Rosanna Hertz in her study of single mothers found that lack of a father led to insecurities for boys who have no one to mirror and have ghostlike fantasies about a father.[11] Other studies, however, find that fatherless boys have no problem establishing their gender identity as male.[12]

I see now that I should have insisted that James, Marcia, and I sit down and talk about our conflicts, clarify our expectations, and work out compromises. Finally, in the summer of 1984, seven years after we moved in, but with increasing tensions between us, we three adults entered therapy together and at the same time I found my own therapist. Marco was three and a half.

In our first joint session, we agreed that we wanted to try to resolve our differences so that we could all stay in the house. But in subsequent therapy meetings, I began to voice my frustrations. Cautiously at first, and then with increasing vigor, my anger came tumbling out—anger at their dominance and my submission, anger at their paternalism for which they implicitly wanted my gratitude. I rejected Marcia's desire that my door always be open to her. As a result, Marcia said that she couldn't live with this degree of tension and was dreaming of other houses. Did she really want to leave? For a while I wasn't sure, but I too began to think that this household would never provide a family for me. Eventually, we all agreed that we wanted to terminate the communal household and at least one party would have to move. Since they had the money to buy me out, they assumed I would move. I refused, because I feared that they would find another single parent or a couple with a child to move in, recreating a community from which I would be excluded. With no partner and disillusioned by the failure to form a collective family, I still craved the community that we had before we had children.

The joint therapy terminated, and more than four months of conflict followed which involved lawyers. Eventually, Marcia and James bought another house, and so did I. The anger and distrust on both sides were too deep to continue a relationship.

The breakup of the communal household was the equivalent of a hostile divorce and was one of the biggest disappointments of my life, more devastating than the failure of any relationship with a man. I had invested my adult identity and commitment in this household that had now fallen apart.

Single parents more recently and in future generations may be more successful. Psychologist Bella DePaulo, in *How We Live Now: Redefining Home and Family in the 21st Century,* interviewed single parents who have found new ways to connect with others who wish to share parenting.[13] This can involve the use of internet sites, like *CoAbode,* for single mothers to find housemates, or *Family by Design* and *Co-Parent,* for men or women who want to find others with whom they can have or raise children. Such searches are not based on dating, romance, or sex, but do necessitate extensive conversations and sharing of expectations, something lacking in my attempt at communal living.

I'm sure that the stress involved in the breakup affected Marco deeply, as did the loss of a relationship with Billy. Throughout his childhood, Marco asked me periodically why we stopped living in that house. I gave vague answers about my incompatibility with Marcia and James. I said I was sorry that we had lost contact with Billy. Was this enough? Probably not. Now I know I should have reassured him that it was not his fault.

Social psychologists have, for many years, demonstrated a correlation between parental divorce and higher rates of behavioral problems in children of divorce, especially during adolescence. The level of family conflict before and after the actual separation or divorce has the most negative impact on the child, as does the loss of contact with a parent or sibling. But most of this research has been done on biological families, where the impact of divorce cannot be separated from parents' personality problems that are genetically influenced and may be passed on as predispositions to their children. Such predispositions may lead to adolescent and adult problems independent of divorce. Behavioral geneticists, seeking to address this issue, designed adoption and twin studies to try to separate the genetic and environmental risk factors when looking at the results of divorce.

The longitudinal Colorado study of adoptive families discussed in previous chapters evaluated adoptive children when they were age twelve, comparing those whose adoptive parents had divorced with the children who still lived in an intact family.[14] Both groups were compared with twelve-year-olds in biological families, divorced and not. On only one measure—substance use—were adopted children in divorced families shown to have much higher levels than those in intact families.[15] This study concluded that an environmental factor (divorce) was important and independent of genetic factors in explaining children's later substance use. The age of the child when the divorce occurred didn't seem to matter.

As I was reading this research, I remembered the adoption study in Iowa that found a strong genetic component in adoptees who had adult problems with addiction. Later, these researchers included environmental factors and discovered that divorce in the adoptive family increased the likelihood of adoptee drug abuse.[16]

Another study, using a (larger) sample of adoptive families from Minnesota with a control group of biological families, found higher rates of delinquency in adolescents in divorced families, whether adoptive or biological.[17] They, too, found that most of this acting out in children of divorce could not be explained by genetics.

None of these studies explain what it is about divorce that makes children vulnerable. Does the stress alter pathways in the brain? Does the stress connected to divorce change the endocrine response to later stress? Or does this environmental stress alter genetic expression, perhaps permanently? All of these possibilities have been proposed, but investigation seeking to link behavior and bodily processes is still in its infancy.[18]

Whatever the reasons, behavioral genetics research links divorce (a defect in nurture) to a greater potential for adolescent and adult addiction. For Marco, genetic and environmental vulnerabilities probably reinforced each other. As with a genetic predisposition, environmental risk factors are not necessarily causal, but they are probably cumulative. Because Marco inherited a genetic susceptibility to substance abuse, environmental stresses increased the likelihood he would face a tendency to drug and alcohol abuse in his teenage and adult years.

Kay and Marco (eighteen months old) at the National Women's Studies
Conference, Humboldt State University, 1982.
Photo from author's collection.

Marco fishing in Yosemite National Park in 1993, age eleven.
Photo from author's collection.

Kay and Marco backpacking in Yosemite.
Photo from author's collection.

Marco and Kay at a family wedding in Rochester, New York, in 2006.
Photo from author's collection.

Marco with his Louisiana nephew in 2010.
Photo from author's collection.

Kay and Marco in August 2019.
Photo by Shelly Hamalian of Lafayette, California.

CHAPTER 5

Childhood Trauma

While dealing with the impending termination of my communal household, I received some disturbing information about Marco's behavior from Jennifer, a colleague and a divorced mother with whom I shared childcare for a few hours on the weekends. Jennifer took me aside one day and told me she found Marco (then three and a half) playing a sex game with her son, Ted (age four). Jennifer, a feminist who was open about sexuality, wasn't upset initially since she knew that age three to four years was when sexual exploration started. She became disturbed, however, when Marco picked up a stick, pointed it at Ted's anus, and asked him to play the "killing game." Jennifer stopped the play, but didn't reprimand Marco or Ted. I was grateful that she didn't cancel future play dates. We were both more watchful when the boys played together. I felt I had to tell my other friends who had kids (all boys) with whom Marco played. No one else saw this kind of play and they were as ignorant as I about what was appropriate or abnormal sexual behavior at this age.

A few weeks later, Marco's preschool notified parents that they had invited a local feminist group to do a presentation to the children on preventing sexual abuse. I approved of such education to teach children the difference between good and bad touching, to instruct them that they could refuse such overtures, and to encourage them to tell an adult about anyone whose touch they didn't like. A week later, Marco watched an E.T. movie on TV that scared him. While I was talking to him about it, he suddenly said:

"I played the killing game with Josie." Josie was one of the black women teachers at his preschool.

Startled, I tried to stay calm as I asked Marco more. He described seeing Josie's vagina, smelling it and her sticking things into him. Although upset, I didn't say anything negative to Marco, but the next morning I went in to talk to the director. She agreed that it seemed probable that Marco had been exposed to inappropriate sexual activity, but she refused to consider that it had happened at her school. I then thought back about his earlier childcare placements.

When Marco was nine months old, I went back to work and placed him in full-time daycare. This proved to be a difficult transition for me. I didn't have adequate knowledge about how to choose a good daycare or how to sense a problem. My circle of friends with babies—most coupled and a few single—was no more informed, for we were the first generation of middle-class white mothers to go to work in large numbers when our children were still infants. As a single mother, I had to work, but even if I had been married or coupled, I would not have given up my career. My married friends, now with children in their late thirties and early forties, felt the same way. The alternatives until age two and a half were family daycare, cooperative childcare, or a private nanny, all with distinct disadvantages. In 1981, no professional daycare centers existed for infants.[1]

Because Marco was too young to communicate his feelings in words, I had to use my intuition about whether a particular daycare was a good placement. For his first four years, I kept changing Marco's type of care, looking for a better fit. Because Marco was so expressive, he often indicated nonverbally that he was unhappy. But why was Marco dissatisfied? Perhaps his temperament—his extroversion and desire for attention— made him a more difficult child for overburdened daycare attendants. Race probably had something to do with it. Even though most of the children in his various placements were white, the caretakers were not. I deliberately chose African American caretakers to avoid raising Marco in an all-white environment. Could these people of color have been stricter with a black child than with my friends' white children? Possibly. Or, maybe, I was overly concerned. I found it difficult to figure out how to raise a child with a temperament different from mine.

When he was two, after he had outgrown the infant daycare and after I found the cooperative childcare too disorganized, I placed Marco in a family daycare in Berkeley run by the Johnsons: a black woman and her adult daughter with some help from an adult son. After three months, Marco came home one day, grabbed one of my belts, and started swishing it around saying, "I'm going to whip you." His play wasn't directed at me, so I wondered what might be happening in daycare.

Mrs. Johnson readily admitted that they threatened to whip children who were misbehaving: "Oh yes, we say that, it's part of our culture, but we would never do it."[2]

I did not believe in trying to control children using violent threats, and as soon as the semester ended, four weeks later, I got Marco out of there.

I felt fortunate that at age two and one half he was admitted into a small, professional preschool owned by the wife of a Berkeley professor. I was pleased that there were a few other black and mixed-race children and two black women teachers, along with one white man. Marco had been in his last two placements for only five months each, and because I thought that all these changes were probably upsetting to a small child, I hoped that Marco could stay there until he was ready for kindergarten. Marcia and James followed my lead and enrolled Billy there as soon as he was old enough. Could sexual abuse have happened there?

Child sexual prevention education, which was initiated by feminists, said adults must take children's disclosures seriously, and I did. Half a year earlier, Marco's playing at whipping turned out to have indicated a reality in the prior daycare. To ignore his words would send a terrible message and might put him in jeopardy. I told the director that I had to remove my son.

Marco's accusations occurred in late 1984 after the widely publicized scandals a year earlier concerning sexual abuse in preschools, like the famous McMartin case and others. There was controversy at the time, and still today, about the veracity of the charges. In those cases, a number of children made accusations, but in my case, no other children or parents came forward. Marcia and James refused to believe anything had happened to Marco at this school, or to their child and others. They informed other parents that they thought I was "overreacting." One mother sent me a note saying she would have acted as I did, but no other parent contacted me or asked for my story. I became more and more isolated, and my anger at James and Marcia increased.

My attention now had to be with Marco. I was numb, unable to cry. I had to hold myself together and rationally decide how to cope. In other situations, a more emotional response might have been better, but here my stiff upper lip was appropriate,

A friend recommended a psychiatrist who specialized in evaluating claims of child sexual abuse. I made an appointment with Dr. Jones for a diagnostic session and I stayed with Marco during the hour appointment. Dr. Jones asked direct, but not leading, questions, and he was warm and empathic. He made an audio recording of the session which he later gave me, an audio record from which I derived much of the detail below. In writing this account thirty years later, I find myself reliving the terrible tension I experienced then. Suffice it to say that I've left out some of the details and provide only highlights of the interview.

Marco, when asked where he played the game with Josie, replied, "In the school office." I didn't say anything, but this seemed improbable to me. The office was in the middle of the small open house with a window that looked out on the yard. Though seldom there when I visited the school, I always had seen teachers and kids going in and out. Part of me still couldn't believe this was happening and I wanted any reason to doubt Marco's story. In retrospect, I wonder if he picked up my doubts.

At various times during the interview, Marco sang a rhyme, "beat them till they're black and blue," which he said was part of the killing game. "I hate this song," he declared. At one point Marco said "Josie wanted me to beat her."

"I beat Josie on her bottom. She liked it."

After these clear assertions, Marco's answers become less precise. When asked where he first played the game or who first played it with him, he said that he didn't know. When asked why it is called the killing game, Marco answered, "That's what it's called." When Dr. Jones asked if anyone said that they would kill him, or kill his mother, Marco wouldn't answer.

At the beginning of the interview, Marco was very clear about the differences between men and women. Men have penises and women have vaginas. He said clearly that Josie was a woman. But near the end of the hour, Marco stated that Josie had a penis and was a man. When pushed by Dr. Jones, Marco asserted that he was telling the truth. At the end of the session, Marco said that ghosts and monsters were involved in the game.

In retrospect, I wonder why Dr. Jones didn't have Marco physically examined. Clinicians today recommend an examination.[3] But I don't know what was common in 1984. At any rate, I can't fault Dr. Jones, for I never told Marco's pediatrician, a woman whom I liked, about the possible abuse or asked her to do an exam. My guilt and shame restrained me, as did my strong desire to minimize what had happened. How could my child have been sexually abused? What was my responsibility?

Dr. Jones said that he didn't know what to make of Marco's interview. He saw evidence of inappropriate sexual knowledge and of trauma, but that seemed to be mixed with fantasy. Especially troubling to him was the inconsistency about whether the perpetrator was a woman or a man. He said that Marco's accusations would not stand up as testimony in court. I was relieved, for I didn't think I could have endured a courtroom, given how tense, upset, and confused I'd become. Dr. Jones advised me to put Marco in therapy. I quickly did so.

At this point, in mid-December 1984, my parents arrived in town to take Marco and me to Hawaii with them for a prearranged vacation. What a relief! I had never been so glad to see them. I gave my parents only a cursory report on my life and I had a wonderful time in Maui, coming back relaxed and fresh. Marco seemed happy too.

I found a new preschool for Marco, one where several friends enrolled their children, and whose director would be especially sensitive and kind to Marco.

Dr. Jones had recommended a young black male psychologist, Tom Harrison, who had a practice specializing in children who were victims of sexual abuse. Marco started seeing Tom for play therapy once a week. I was relieved when after a few weeks of therapy, any references to the killing game stopped and there were no more attempts to play sex games with other kids.

Marco stayed in therapy with Tom for the next six months and then saw him occasionally for another half year. Tom was warm and friendly, and Marco especially liked Tom's dog, which he often brought to the office. I thought Tom was a good role model, even though he seemed inexperienced and had never seen a patient as young as Marco.

Tom reported to me that Marco had recanted his story about Josie, saying he just made it up in his head and refusing, or unable, to talk more about the game. In retrospect, I wish I had asked Tom why Marco might have focused on a black woman. Tom thought Marco had been sexually abused, and like Dr. Jones, Tom told me that Marco demonstrated sexual knowledge and behavior improbable for his age, that he had a negative self-image, and that he expressed a lot of anger. When I asked why Marco couldn't talk about it, Tom didn't have a ready answer for Marco's silence. Maybe the abuse had happened so young that he couldn't verbalize it, or he was afraid to talk, or he now wanted to forget about it. My therapist, a middle-aged woman as experienced as Tom was not, had been completely surprised by Marco's accusations, but she took them seriously, didn't think I had overreacted, and supported my action of removing Marco from the school. She told me that forgetting a negative experience was probably important for Marco's normal psychological development

I also wanted to forget. I began to hope that maybe it had all been a fantasy. I could teach my Women's Studies students about child sexual abuse, and I knew it happened in all social classes, but I never thought it would be an issue I personally had to confront.

Marco loved the new preschool, and he seemed to be his old happy, extroverted self. Then three months after he arrived, the child sexual abuse prevention project came to this school to do a similar, but longer, three-day presentation. After the first day, Marco said he hated the skits, and didn't want to go to school. I had to go to work and had no choice but to take him back to school the next day, where, for the first time in several years, he cried when I left. When I returned to pick him up, the director told me that Marco had initiated sexual play that day. In therapy that week, Tom reported that Marco reverted to being the "bad guy." But as soon as the three-day education program ended, Marco returned to normal play and again became the "good guy" in play therapy.

What was going on? Could he have acted out because he thought that these sexual abuse presentations would mean he would be taken out of this school as he was from the last one? But that didn't explain his sexual behavior. In 1985, I hadn't looked for research about childhood sexuality and childhood memory and what that research might tell me about the veracity of Marco's accusations. I was a single parent now living on my own, and I had all I could do to keep my work and life together and to try to provide Marco with a safe and calm environment. After Marco's reaction to the second child assault prevention presentation, I began the difficult process of accepting that in all probability, Marco had been sexually abused.

I obsessed about his childcare teachers and babysitters and even about all the people coming in and out of the communal household.[4] Finally, I had to accept that I would never know what, when, or where it happened. I believed then and still believe that I had done the right thing in listening to him, leaving the communal household, changing his preschool, and getting him into therapy.[5] But it had taken a toll on us both.

Had my being a single parent made Marco more vulnerable? Even if I had had a husband or male partner, I would have worked, put my child in daycare, and used occasional babysitters. Researchers have found, however, that child sexual abusers may target children of single mothers, seeing them as overworked and less likely to investigate. Or they see a child of a single parent as more emotionally needy, which often may be true. Most sexual abusers are men, and most single parents are women, making it likely that a boy or girl raised by a single mother lacks a father.[6] In addition, if I had been coupled, I might not have lived in a communal household, a setting that could have been an additional risk factor for Marco.[7]

Years later, I decided to investigate academic research on child sexual abuse. If I had looked in the mid-1980s, I would not have found much research on normal childhood sexuality, on sexual abuse of young children, or on childhood memory. From 1990 on, however, there has been an impressive body of research studies and clinical data that has helped me interpret Marco's situation. I discovered that it is still difficult to diagnose sexual abuse in children less than five years old.

Researchers have tried, and are still trying, to answer the question I first asked: What is normal childhood sexual behavior and how does that differ from the behavior of children who have been the objects of inappropriate sexual attention from adults or from older children or teens? It is difficult to answer this question because it is hard to study sexuality in a nonclinical population of young children. Most parents will not give consent to have young children be part of such a study.[8]

The best study I found designed to assess normal versus abnormal childhood sexuality was published in *Pediatrics* in 1998.[9] The findings for 287 boys between ages two and five were that while some sexual behavior was very common (e.g., 60 percent touched sex organs at home), behavior like Marco's when he was not yet four was very rare. Only between 0.4 percent and 4.6 percent of preschool boys initiated the kind of sex play Marco had exhibited.[10]

In *Assessing Allegations of Sexual Abuse in Preschool Children: Understanding Small Voices*, Sandra K. Hewitt, a PhD psychologist who had worked for twenty-two years clinically assessing sexually abused children, cited her own research and that of others to conclude that "sexualized behaviors (e.g., oral sexual contact, direct genital contact in a sexual way, etc.) are rare in a population of non-abused preschoolers."[11] Many cases of sexual abuse, Hewitt reported, are first discovered from inappropriate sexualized play; she found that preschoolers will play very graphic reproductions of their own or witnessed sexual abuse, even when their immature language skills do not lead to a coherent narrative. She found that even when an account of sexual abuse was validated, the child could include fantasy elements. The use of fantasy was most common among children known to have suffered severe abuse. Hewitt concluded that children could use fantasy in their reports of abuse to master anxiety, to avoid blame, to deny victimization, and/or because of cognitive immaturity.[12]

Marco's response was much like that of a child in one of Hewitt's case studies, a four-year-old child who described specific instances of abuse that happened a year earlier, but who combined those details with confusing, questionable, or unreliable information. This child, like Marco, could not give a coherent narrative account of what happened and wouldn't answer some questions.

Hewitt brings up the question of whether children, as they become more verbal between ages three and four, can report abuse that happened earlier. She found several examples in her clinical practice where this was the case. Research I found on childhood memory validates Hewitt's observation. Psychologists began only in the mid- to late 1980s to study childhood memory directly rather than rely on adult recollection. Because the vast majority of adults cannot remember events before they are 3.5 years old, psychologists had previously assumed that infants and toddlers didn't have preverbal memories. Research in the past twenty-five years, however, solidly demonstrates that babies as young as one year have nonverbal recall and by two years of age, nonverbal memories are reliable. By age three and a half, children can remember what happened more than a year earlier, especially when prompted by questions, although their verbal report will not be as detailed or specific as that of older children.[13] But preschool children are more susceptible to leading and suggestive questioning than older children or adults, setting up another barrier to documenting sexual abuse in young children.[14]

Scientists have found that verbal memories can be forgotten, and children forget faster than adults. Both negative and pleasant events can be lost from memory. The most convincing evidence that childhood sexual abuse can happen and be completely forgotten is a sociological study by Linda M. Williams. Williams located and interviewed 136 adult women whose childhood sexual abuse was documented in hospital and police records. Thirty-eight percent of these women between ages eighteen and thirty-one didn't remember the abuse. For those who had been three or under when the abuse was reported, 55 percent of the adults had no memory of it.[15]

All of this research reinforced my belief that Marco at an early age had been exposed to some sort of inappropriate sexuality, and that the prevention programs had helped him reveal it. His only partially articulated and changing verbal stories at age three years and ten months, and his sexually acting out, are consistent with the idea that the sexual exposure could have happened when he was younger—two to three years old—and that the prompt of the programs could have released the memories, some of which were nonverbal.[16] He could have transposed preverbal and barely verbal memories onto the daycare teacher or his accusations of her could have been true.

Because Marco was such a happy and accomplished boy from ages four until thirteen, I never considered that this early trauma might have an impact on him as a teenager or as an adult. Now I find that behavioral geneticists have investigated the possible biological links between child sexual abuse and adult psychopathology.[17] Using subjects from the Iowa Adoption Project, researchers wanted to find out whether child sexual abuse has an effect independent of genetics and whether a genetic predisposition to substance abuse and other forms of pathology increases the impact of child sexual abuse. Both proved to be true. The researchers concluded that if an adoptee had birth parents with drug or alcohol problems, a history of child sexual abuse would further increase the probability of substance abuse for the adult adoptee. Thus, genes and environment interact to generate a greater impact.[18]

Steven Beach and his colleagues used data from the Iowa adoption study to investigate how child sexual abuse predisposes to adult substance abuse and psychological problems even when the individual had no genetic risk factors. They discovered this could happen through an epigenetic phenomenon. *Epigenetics* refers to changes in the cell structure around the gene that can alter genetic expression, including turning genes on and off. These changes can be long-lasting without changing the DNA structure of the gene itself.[19] Thus, one reason that early trauma can have enduring psychological effects is that it can induce epigenetic alterations that cause long-term changes in gene regulation for the child as s/he matures.[20] Beach and his colleagues found evidence of epigenetic change in girls who had been sexually abused and who then exhibited antisocial personality disorder as adults.[21] Other researchers have looked at the impact of the stress of child abuse on brain development and hormonal functioning.[22]

Research on these complex chemical processes in the body is in an early stage and the details are beyond the understanding of those of us who are not scientists. But I accept the strong possibility that Marco's early childhood trauma had long-term psychological consequences.

◆ ◆ ◆ ◆ ◆

Throughout his childhood and into his adolescence, Marco never made any reference to sexual abuse or asked me about it. When I mentioned that he had some bad experiences in daycare when he was young, Marco didn't question me and gave no indication that he remembered this history. Linda Meyer Williams's research demonstrates that being sexually abused as a child and having no memory of that as an adult is entirely possible.

When he was nineteen, I decided to tell Marco the story I've recounted here in a therapist's office and to give him the audiotape of the interview with Dr. Jones. Marco and I had started seeing a therapist together to deal with our increasing conflicts about Marco's life choices. I called the therapist beforehand to tell him what I wanted to do. He replied that he wanted to ask Marco's consent at the beginning of the session; Marco readily agreed.

As I told the story, I noticed how Marco was completely focused on what I was saying. When I finished my narrative, Marco made two comments.

"This helps me understand some things about myself," he said.

Then he went on to look at me and say, "It must have been so hard for you."

I began to cry. A burden I had carried for fifteen years had been lifted. I had transferred the story and what it might mean to my adult son, a son who had empathy for, and did not blame, his mother. However, my relief was never complete. I now believe that Marco's problems as an adult stem, not only from a biological inheritance from his birth parents, but also from the instability of his home environment as a young child and from too early an exposure to adult sexuality. Even though I don't know how I could have protected Marco from the sexual abuse, I still grieve for my inability to better protect my tiny boy.

CHAPTER 6

The Good Times

I look back on the years when Marco was five to twelve as an idyllic time when we were both happy. My emotional attachment to him, present from the start, deepened during these years of middle childhood, which were so much less conflicted than those in his early life. During this time, in choosing my son's environment, I tried to respond to his temperament and interests, while also acting on my values. Through our life together, we cemented a strong bond that lasted through the upheaval of his adolescence and early adulthood.

The house I bought after the breakup of the communal household when Marco was four was in a mixed-race neighborhood in the flats of Berkeley, a mile south of the university and downtown. As a professor with a single income, and having the profit from selling the communal house, I could, in the mid-1980s, afford to buy my own small house.

The new neighborhood appealed to me, not only for its racial diversity and convenience, but also because of its variety of styles and sizes of houses, from large Victorians to stucco cottages to brown-shingle bungalows like mine. From my house, I walked a few blocks to the Ashby BART (Bay Area Rapid Transit) station for easy access to San Francisco. I walked to the Berkeley Bowl Market, the biggest produce market in the city. I could drive a few minutes to year-round farmers' markets on four days a week. I walked to the supermarket on Telegraph and to the trendy restaurants and stores on College Avenue. For exercise, I walked through the upscale neighborhoods and hills of the Elmwood and Rockridge districts, and I drove five minutes to the YMCA gym downtown. Tilden, Redwood, and other East Bay parks in the hills offered great hiking with only a fifteen-minute drive. As he became older, Marco made friends—white, black, Asian, and mixed-race—on our block and in the neighborhood.

We lived on the top two floors of a compact, three-story brown-shingle house built in the 1930s. The main floor consisted of a living room with a working fireplace connected to a dining room with built-in cabinets, the top doors of which were fitted with leaded glass. Behind was a kitchen and breakfast room. Back there, sun flooded in from the southern exposure, but because the front two rooms were somewhat dark, despite plenty of windows, I painted the ceilings, all the woodwork and fireplace white, offset with colorful walls. Throughout the house were lovely oak floors. Upstairs, I gave Marco the larger front bedroom with a view of the Berkeley hills, because the middle bedroom, along with my sunny back study and the bathroom, formed a unit for me. The property included two rental units—a studio apartment on the ground floor and a studio cottage in the back yard, over which I built a deck with access from my kitchen. The rental units provided income to offset the mortgage payments.

This cozy house proved to be small enough for intimacy but large enough to assure me some privacy. I loved living in this house with Marco as a child, disproving my fears of single parenthood. It was so much easier than trying to build a communal family. I still live here more than thirty years later. Marco says he never wants me to sell the house.

Learning about my adopted child was an ongoing process of discovery and interaction. I liked Marco's extroverted personality and I reacted positively. I liked that he always had lots of friends. I, as well as his cousins and friends, marveled when Marco at age seven rode a unicycle in a clown suit and wig, making jokes as he juggled. We loved his imitations of the nerdy TV character Urkel, from the sitcom *Family Matters*. Despite his joking, Marco was never out of control and never had an accident, even when he took up skateboarding and snowboarding, interests I didn't share.

Marco was enthusiastic about participating in the outdoor activities I loved—hiking and backpacking. I encouraged his abilities in theater, music, and the arts—abilities I valued, but didn't share. One of his drawings from elementary school, an abstract still-life in charcoal with a watercolor wash, so perfect in its strong lines and composition, hung in my bathroom for many years. I still have the handsome old oak piano I bought for him when, at age five, he announced with determination that he wanted to learn to play. I always wished I had been given piano lessons as a child and I wanted to do better by Marco.

Encouraging his interests was important to me, because my mother had denigrated mine.

When I was an elementary school student, she scolded, "Get your nose out of that book."

"Classical music—why would you want to listen to that junk?"

Our house had few books, but from an early age, I loved the public library. I first heard classical music at the home of one of my elementary school friends, Laura Wolfowitz, whose younger brother Paul many years later became the Republican secretary of state.

I never let go of my love of reading or music, something that differentiated me from my parents and siblings. But my mother's anxiety and her lack of acceptance of my specific interests and character led me to withdraw emotionally. In college and afterward, when I wanted to talk with her about problems with boyfriends, she cut me off and didn't want to listen. Because of this failure in her nurturing, I disliked her. I wanted to be a different mother.

Marco had other traits I valued that were different from mine. I was warm, but he was cuddly, much less physically reserved than I, and he was more emotionally astute about other people's moods and desires. I loved his openness. Even though I found it hard at times to hear about situations that made him unhappy, embarrassed, or disappointed, it was wonderful that he could share them, along with positive experiences, in a way I had never been able to do with either of my parents. I responded to his love of animals and became attached to our two cats and two dogs, even though I could have done without the hamsters, rat, snakes, and chickens I let him bring into our home.

Even in those happy years, raising a black/biracial son as a single, white mother was harder than I had imagined. Working full-time with no extended family nearby, I tried to build a friendship network and a community to help me parent, endeavors which sometimes conflicted with my more abstract desire to find a multiracial environment for Marco. My friends and their children gave Marco a sense of family, but it was a white family. My attempt to find male role models for him led most often to white men. My desire for an environment that embodied my values sometimes was in conflict with looking for a multiracial one. As a feminist I wanted my son to attend a school that actively sought to challenge traditional gender roles, but this clashed with my desire for a multiracial learning community, where ideas about gender seemed more traditional. In a predominantly white school, I was happy that Marco bonded with the only other black/biracial boy in his class. Thus, my nurturing often involved compromises, some of which worked out better than others. My situation had unique characteristics, but also many similarities with the issues faced by all parents.

One major decision in my search for the best environment for Marco was choosing an elementary school. Although everyone in my family had a public school education, and I believed in public funding for education, I decided to send Marco to a small, private, progressive elementary school in Berkeley, the New School. I had to refinance my mortgage to pay for private school, but doing so was a good decision. In sending Marco to a small school, where I knew some of the teachers, I was motivated by a desire to protect him after the turmoil caused by my "divorce" from the communal household and by his sexual abuse. I'd met a number of the New School teachers when I had volunteered for a week (before I adopted) in a multiracial summer camp for six- to twelve-year-olds. I was impressed by these teachers' skills with children and by their personal commitment to children from varying race and class backgrounds. These values were expressed in the creative, multicultural, and multiracial curriculum of the New School, and in its practices such as giving written evaluations in place of grades.

Despite the curriculum that emphasized diversity, all the full-time teachers were white. But at least I knew that the school would be less likely to consciously discriminate against a black student. I liked that almost half the teachers were men, and I especially liked a mild-mannered Ghanaian man who worked in the afterschool program and coached a soccer team. He was a good role model for Marco, as were other white and black male mentors he met outside the school. There was Dave, one of the leaders of the Young Explorers program for hiking and backpacking; Michael, his piano teacher; Russell, an African American man who coordinated a summer musical theater program; and my friend Robert, who helped Marco learn to fish.

Like the other middle-class single mothers that I and others have studied, I sought to compensate for the absence of a father by bringing male role models into Marco's life.[1] The examples above provided positive experiences for Marco. They didn't really replace a father, but their influence may help explain why children raised by middle-class, educated single mothers do not display deficits as adults. Outcomes are no different than those for children raised by two parents.[2] But my earlier book, *The New Single Woman,* documents the negative experience I and another single mother had with trying to find father substitutes through the Big Brother organization.[3]

Marco received an excellent education at the New School and won special praise for his writing and his artwork. Five out of twenty students in Marco's class were nonwhite, but only one other of these predominantly mixed-race children was a black boy. At the local public school, Marco would have had more black classmates and teachers, but he would have been exposed earlier to a male subculture that I thought might put him at risk for conflict with the police. A few years ago, I discovered an ethnographic study of a public elementary school for grades 4–6 in the Berkeley flats, a study conducted from 1990 to 1993, the very years when Marco was in these grades. African American sociologist Ann Arnett Ferguson followed twenty black boys in fifth and sixth grade for two years to produce *Bad Boys: Public Schools in the Making of Black Masculinity.* Children in our neighborhood went to this school, so if he had gone to a public school, Marco could have been one of her subjects. Because of his skin color, Marco could have been one of the boys labeled by this school at an early age as "at-risk, unsalvageable, or bound for jail." In this public school, Marco would have been exposed earlier to the male ethos of fighting to prove oneself, a cultural norm especially prevalent in poor black families unable to adequately protect their children.[4]

Reading this book made me realize that I would have had a much harder time being a parent if Marco had gone to this public school, reaffirming my decision to send him to a private one. The New School protected me from having to deal directly with racial and class inequalities. If Marco had attended public school, maybe I would have been better prepared to deal with these issues when they arose in Marco's teenage years. But Marco could have been damaged by an earlier exposure to this more challenging environment.

The anecdotes that follow illustrate my complicated attempts to deal with race, class, and gender during these years for a black son who had neither a black nor a male parent.

I hadn't consciously focused on nonsexist teaching when I chose a school, but I was pleased when I found that the New School encouraged a wide range of behavior in boys. Marco announced in his second month of kindergarten that he was taking his black male doll Martin (named after Martin Luther King) in his stroller for "show and tell." I was happy Marco was not aggressive, but I feared that a boy's interest in a doll might lead to bullying. As a single mother, I also worried it might reflect badly on me, that Marco would be seen as too feminine because of the lack of a father. But I squashed such thoughts and consented.

"Marco," I said, "some kids might tease you about having a doll."

"No one will do that," he replied.

Does such confidence come to a child acting on his own feelings and interests rather than because of parental- or peer-imposed rules?[5] Later, my own research and that of others demonstrated that single mothers who were feminists raised children who were more flexible in their behaviors than those prescribed by traditional ideals of masculinity and femininity. Arlie Hochschild in The Second Shift, her study of parents who did and did not share work in the home, found that the fathers who shared the most had been raised by single mothers or by married women whose husbands were mainly absent.[6] The next day the kindergarten teacher pulled me aside and said it was wonderful that Marco brought the doll. "Even the boys pushed it around," she said, "and no one laughed."

In the women's studies classes I taught, I advocated for the acceptance of homosexuality, and I had gay colleagues with whom I was comfortable, but it was one thing to talk about these issues intellectually and another to deal with my inexperience in relation to my own son. One day Marco came home from first grade and said, "I want to get my ear pierced and wear an earring." He explained, "A boy in the sixth grade has one and it is so cool." For several months, I gave lame excuses. He was too young. It would hurt. I didn't know where to get it done. But he persisted, found out it cost eight dollars, and got the name of a good store. I discussed my ambivalence with friends. Not only was I afraid he had too much female, and not enough male, influence in his life, but I worried that my son would be labeled a homosexual. A friend asked me if he would pierce his left or right ear.

"Does it make a difference?" I asked. I had no idea that gay men signal their homosexuality by an earring in the right ear.

When I asked him, Marco said he would pierce his left ear since he thought he was straight. It turned out he was right, but at the time I didn't give this much credence. At age six, could he know his own sexual orientation? Maybe he had come in contact with a gay father of one of the children at school or saw something on TV. But I didn't take his answer seriously.

I finally decided that piercing his ear was not going to make him gay. "If this is a way to express his feeling about being a homosexual," I thought, "then my forbidding it is just going to make him feel bad about himself."

I remembered the negative impact of my mother's insensitivity to my preferences in appearance and dress. In particular, she denigrated my hair: "Your hair is a mess. . . . You need to get it cut and styled."

Not until I was away from home, and well into my twenties, did I let my hair grow long. Only then, after much praise from friends and strangers, did I realize that my thick auburn hair was my best physical feature.

My mother sought to dress me and my (much younger) sister according to her understanding of middle-class girlhood. When I look at photos of myself from age five through age twelve, I see an unsmiling, pudgy girl with thick glasses and a bad haircut. One picture at Easter shows me wearing an ugly blue suit, bobby socks, white gloves, and a pillbox hat on my head. I remember feeling as awkward and uncomfortable as I looked. Not until I was a teenager did I assert my own taste in clothes, which my mother reluctantly accepted, but not without often making negative comments.

I did not want to replicate my mother's negativity. I wanted a son who was happy and self-confident, whatever his sexual orientation. He pierced his ear and put in a silver stud. When people asked me why I let him do it, I answered, "It cost only eight dollars and the hole will close if he changes his mind." Soon, many boys in first grade wanted an earring, but no other parents permitted it. I didn't know if they disapproved because they were homophobic, because they held conventional ideas about masculinity, or simply didn't think children should have pierced ears.

"She doesn't like my earring," Marco said about one of my friends with a son his age. "How do you know?" I asked.

"Because she's asked me about five times why I did it and never accepts my answer."

Negative comments didn't undermine his happiness, however. Soon afterward he saw a multicolored baseball cap he liked. Every day he wore this fancy hat, the bill to the opposite side of his earring. At the end of first grade, his teacher commented that acquiring the earring and hat marked a change in Marco's behavior. "He is more willing to participate in class and more self-confident," she said.

When he was in third grade, the New School introduced an eight-week program in five performing arts. His teacher told me that Marco's hand shot straight up, with no hesitation, for his choice—jazz dance. However, only one other boy attended the first class drawn from all the grades in the school, and after the second class, Marco came home dismayed that the other boy had dropped out. I suggested that he recruit other boys, but he said this wouldn't work. Still, he loved the class and wasn't interested in the other choices. "O.K.," I said, "I'll speak to the director and if you decide later you want to change, I'm sure she'll let you." Then I asked the male third-grade teacher to give him support. Marco continued in jazz dance and every week he showed me the new steps he had learned.

A crisis arose when he learned that each class would perform publicly. He just couldn't do it. His teacher, along with me and the director of the arts program, encouraged him to perform, but we left the decision up to him. Two days before the show he announced he would dance. His biggest anxiety, it turned out, was that he was one of the youngest in the class. He thought he couldn't dance as well as the others, many of whom were older girls who had had previous dance lessons. Once he felt competent enough as a dancer, he was ready to perform. He danced beautifully. I had no ambivalence then. I was proud that my son was the only boy who would dance like that. I glowed when other parents came up to Marco and me to praise his performance. I'm sure he felt good too.

I was always on the lookout to support Marco's abilities, and several months later in a video store I spotted a cassette of the film *Tap* starring dancers Gregory Hines, Savion Glover, and many of the famous old black tappers. I brought it home, not knowing whether Marco would be interested. He loved it. For weeks he and a white, male friend went around the house and schoolyard making up tap dances. Number one on his Christmas list was tap shoes. I got him the shoes and spent several days on the phone finding a tap class for his age group that included at least one other boy, as his friend had lost interest. On the first day of class, the teacher said to Marco: "You look just like Gregory Hines, earring and all."

All at once it clicked. The earring and his interest in dance might have nothing to do with femininity or sexual orientation. Whatever his ultimate interest, talent, or sexual identity, wearing an earring and dancing were *not* attempts to be feminine like his mom, but a way to *separate* from me and develop his own identity as a black male. The most positive images of black men in our culture focus on music, dance, avant-garde style, and, of course, sports. A black son growing up in a white family, I realized, relies on cultural images. If I had adhered to a narrow ideal of white, middle-class masculinity and rejected his desire for an earring or his interest in dance, I might have thwarted his masculinity and his racial identity.

Marco's best friend was Nick, the one other black/biracial boy in his class at the New School. Nick, too, had a white single mother, Lisa, quite a bit younger than I, who was a free spirit and lots of fun. Her high-achieving son was on a scholarship. Marco loved hanging out with

Lisa and Nick. Lisa, although from an upper-middle-class family, was a college dropout who worked as a secretary, drove a rickety van, and identified as a hippie. Nick often came to our house to play and go biking or to the beach with us, while Marco sometimes went to their house and once joined them on a short vacation farther north in California.

Because she sent her son to the New School, and because I thought it was healthy for Marco to have a good friend who was black, I was less concerned than I might otherwise have been about all the time Marco spent with a mother whom I didn't want as a friend. She was a dropout and a counterculture hippy; I was a professor and interested in political and social change. Though I sensed she used drugs, it never occurred to me that she might offer them to the boys. Only years later did I find out that Lisa gave Marco his first puff of marijuana when he was eight and probably many more after that. Had I known, protecting him from exposure to drugs would have overridden my desire that he have a black-identified friend.

Behavioral geneticists have found that children actively build their own worlds. Their choice of friends, activities, and peer groups seems to be genetically influenced.[7] Children gravitate toward environments that evoke responses from others which allow them to actualize genetic capacities with which they are born. Perhaps Marco's attraction to Nick was not only because of their shared experience of being black boys raised by white mothers, but because of his affinity to Nick's hippie mom and her drug use, attributes which were closer to those of his then-unknown birth parents. Realizing this only after reading many studies makes me wish I could have protected him.

Although the New School didn't protect Marco from exposure to drugs, and didn't give him enough of a multiracial environment, it provided a supportive community for me, a single mother without family in California. Because the school was located near downtown Berkeley, and not in the hills like most other private schools, it attracted many middle-class, but few very rich, families. I fit in with these liberal professionals, and there were enough single mothers to make me comfortable. This community made me happy, and I think Marco picked up on how I valued friendship and social networks. But Marco has maintained no connection to his elementary school classmates and by high school had lost touch even with his friend Nick. However, he made new friends easily and gravitated as a teenager and as an adult to a very different community than I would have chosen for him.

More than twenty years later, I still run into and chat with parents from Marco's class and a few of the teachers. These ties have become attenuated, because it is hard for me to hear about other sons' and daughters' achievements and be forced to say something, however vague, about Marco.

The New School community overlapped with my friendship network, important to me because it took Marco and me outside a narrow mother-son bond and served as a substitute for family support. At the school, I made a new friend, a single mother a bit younger than me who was a graduate student at Berkeley and who had a biracial Filipino/white daughter in Marco's class. My attempt at a multiracial friendship with her and her daughter didn't go very far, however, because our kids were too different. At the same time, I strengthened my friendship with Katrina, a white single mother whose son Kevin was a grade ahead of Marco.

I had become friends with Katrina, a research scientist, a number of years before we both became single mothers. We bonded through our mutual love of hiking. As single women in our thirties, we hiked together locally on weekends and then started taking backpacking trips together in the Sierras, the Grand Tetons, and Hawaii. We had compatible energy levels and some similarities in temperament. Katrina was three years younger than I and gave birth to her white son, Kevin, three weeks before I adopted Marco. In both cases, choice was involved. Katrina chose not to stay with her boyfriend at the time, but to have the baby and to raise him as a single mother.

After my communal household broke up, the four of us spent more and more time together. When the boys were from age four to age ten, Katrina and I took them on a trip together every summer to the California Sierras, camping or staying in tent cabins in Tuolumne Meadows in Yosemite National Park or going to a family camp. Like most family vacations, these trips were not tension free, as Katrina and I sometimes disagreed about child-rearing techniques. I saw her as too strict and judgmental; she, I think, found me too loose and unsure about how to raise my son. But the fun and camaraderie we experienced on these trips overrode our differences.

Katrina's brother, his wife, and their two sons, who were much older than Kevin, also lived in Berkeley. They became an alternative family for Marco. Especially important was Katrina's nephew who often babysat for both boys. Marco stayed with Katrina and Kevin when I had to travel for work or when I wanted to get away for a short vacation. When Katrina traveled, however, Kevin stayed with her brother's family. So ours was not a completely reciprocal relationship, sometimes making me feel inadequate and reinforcing my sadness at not having family nearby for Marco. But both Marco and I benefited from Katrina's generosity and her commitment to the boys' relationship.

When the boys were about ten, Katrina began to withdraw from me, while still encouraging Kevin's friendship with Marco. Hurt and confused, I suggested we see a therapist together. Katrina agreed. She remembers that we talked about several topics, but what I recall is Katrina explaining that she wanted to find a male partner. She believed all the time we spent together was preventing that. I wasn't looking for a partner and saw Katrina only as a friend, someone with whom I was building a family-like relationship. I saw her desire as real, so even though I felt a loss, I withdrew some without ending the friendship. Katrina soon met a man, and although they never lived together, they were a couple. Katrina and her partner now invited Marco on camping trips with them. I was hurt by my exclusion, but I always encouraged Marco to go, while I found other single women with whom to travel when Marco was with them.

The boyfriend had older children, and Katrina was not looking for a father for Kevin. Around this time, he reconnected with his birth father, who lived nearby, and with his older half-brothers. Probably as a result of these changes in Katrina's and Kevin's lives, Marco started urging me to get a boyfriend so he would have a father. I told him that I didn't have time to look for a partner since my work, friends, and he took up all my emotional space. I was happy this way. I said that if I had a partner I would have less time for him. I said that many men my age (fifty) already had children or weren't interested. I only learned later that children in families with a stepfather had more problems than those living with only their mother.[8]

Kevin and Marco, despite their distinct temperaments and interests, retained a sibling-like relationship into their twenties. Marco took the bus to Kevin's college graduation in the Midwest, and they then drove back to Berkeley together. They still see each other occasionally, and Katrina and I are still friends. We participate in a women's walking group and celebrate our birthdays together. The friendships between Katrina and me, and Kevin and Marco, didn't build a family as I had hoped, but did result in significant relationships for both Marco and me.

Despite friendship and community support, single parenting was hard and I wasn't always effective, perhaps making my situation little different from that of any parent. There were times when I was tired and impatient, especially on days which included a long commute. I wish I'd had some guidance in parenting. As an adult, Marco told me I should have been stricter with him. I was unable to provide needed discipline partly due to the difficulty for any single parent of being caring and yet enforcing rules, and partly because of my temperament and values. With friends, colleagues, and students, I wanted to be seen as "nice," not as someone who was tough or demanding. I didn't think about how children developed self-discipline, and never worked out my own ideas about how to be strict and caring.

Because I believed nurture was all important, I thought erroneously that a child just picks up the parent's habits. I was organized, so Marco would be too. I worked hard, so Marco would too. I liked to cook, so he would be a good cook. Only the latter turned out to be true. I gave no thought to any impact of Marco's genetic heritage.

I made excuses for Marco. Didn't all kids lie sometimes? When we visited my prosperous cousin and her doctor husband on the East Coast, couldn't jealousy explain why nine-year-old Marco stole some coins from her son's collection? What if he dropped one activity to try something new? Wasn't it a blessing to be good at so many things? When he dropped piano and took up the saxophone, wasn't playing sax in the middle-school band better suited to his extroverted nature? I presumed that as Marco matured he would decide what interests he most liked and then he would pursue them in an organized way, as I had.

I didn't think I should, nor did I want to, impose my academic interests and temperament on him. I would not be like my parents, who had a plan for my life that I had to fight. When I told my parents at age seventeen that I wanted a career, they decided that home economics would parallel that of my father in animal science. I knew this wasn't right for me, but since I knew no woman whose life I could emulate, I went along and spent my first college year (1958–59) at Cornell in the College of Home Economics (now Human Ecology). I hated it. I liked the domestic life my mother provided, but this was not the career to which I aspired. At eighteen, planning an efficient kitchen or learning why biscuits rise was not the intellectual life I craved.

So in my sophomore year, with my parents' reluctant consent, I transferred to the College of Arts and Sciences and became a chemistry major. I had loved my first-year chemistry course with a famous professor and science was more acceptable to my practical parents than the humanities or social science. By my senior year, I knew I had made another mistake. Hating lab work, finding science theory too abstract, and deciding medical school was not for me, I knew I wanted to go on to graduate school, but I didn't yet know my career goal. Could I be a professor? My parents had encouraged me to be a teacher, but not a college teacher, and I didn't know any women faculty. I applied to graduate school in international and comparative education, inspired by the summer I spent in Honduras on a Cornell student project and by my two teenage years of living in the Philippines. (My father was part of a team of agricultural professors from Cornell who, in the mid-1950s, helped to rebuild the University of the Philippines College of Agriculture.) Maybe I could have some Peace Corps–related career.

After receiving my master's degree in education, I switched again, this time to sociology. All of these alterations in degree programs didn't indicate frivolity, but reflected my seriousness. I was now committed to becoming a professor, but to succeed I had to find the right field. Finally, I had. Sociology suited my broad and changing interests.

I thought Marco would follow the same pattern of deciding which of his many interests to follow toward a career or settled work life, and without my interference. Wasn't my approach even more important for an adopted youth? Later, in his mid-twenties, Marco said he wished I had a family business he could have entered, indicating he didn't like choices.

Only later, through my study of behavioral genetics, did I think about how my nurturing might be related to Marco's troubled teenage and adult life. All behavioral geneticists believe that parents and the family environment make a difference, but they question how much of a difference. To find out why children in the same family develop differently, behavioral geneticists study children in middle-class adoptive families. If there is good enough parenting and an average environment for child development, behavioral geneticists have concluded that whether parents are strict or lenient, involved a lot or a little, doesn't make much difference in how the child turns out as an adult. They agree that the effect of genes is more important than being raised by particular parents. Behavioral geneticists postulate that genetic inheritance explains 40 to 50 percent of the difference between children raised in the same family. The environment in the household and neighborhood appears to account for only 10 to 25 percent of the variance between siblings in personality, cognitive abilities, and psychological problems.[9] The remainder is unexplained in scientific studies, which have to allow for chance and for the limitations in tools to measure complex components of human development. Within the 25 percent, I wanted to figure out what difference I made as a parent.

Temperament, personality, and intelligence, all of which have a significant genetic component, affect how parents, family members, and other caretakers respond to the child.[10] The parents' genes (completely different from the adopted child's) also impact the interaction. This point is made by Eleanor Maccoby, a well-known developmental psychologist, who in 2000 came to accept the findings of behavioral genetics as relevant to the study of child development, while other psychologists at that time still rejected them.[11]

This behavioral genetics research led me to conclude that my liabilities as a parent, including being a single parent, were not grievous enough to explain the negative aspects of Marco's life as an adult. Better nurturing in his middle childhood probably would not have saved Marco from later problems, but more recent behavioral genetics studies have shown me that aspects of my parenting may have made a positive difference to Marco's development.

Starting in 2002, the Early Growth and Development Study followed adopted children, their adoptive families, and their birth families from infancy onward.[12] This project, unlike much of the earlier research, monitors and continues to follow the birth parents as well as the adoptive families. One of its publications looks at how parenting styles affect adopted babies from different types of birth parents.[13] This study separated birth parents into high genetic risk versus low genetic risk based on interviews with them at three to six months postpartum. Birth parents were judged to provide high genetic risk based on three criteria: use of alcohol, tobacco, marijuana, and other drugs; engagement in antisocial (i.e., criminal) behaviors; and incidence of depression or anxiety. Certainly, Marco's birth parents' profiles put him at high genetic risk. The next part of the study observed adoptive mothers interacting with their toddlers at eighteen months and judged them as to the extent the mothers structured their child's play and cleanup.[14] According to the criteria of this study, my parenting provided structure. In their experimental situation, I would have given directives for an action or requested a behavior change. But providing structure has a different impact on young children depending on their genetic risk.

The researchers found that children with high genetic risk had fewer problems when they had structured parenting, like mine, while such parenting increased the problems for toddlers whose birth parents had few risk factors. Structured guidance by mothers provided a buffering effect on the deviant or aggressive behavior of toddlers at high genetic risk. Despite my mother's deficiencies in psychological nurturing, she provided a structured and orderly home life with regular times for good meals together, homework, play, and bedtime. My choice to replicate the organized way I was raised may well have provided the structure Marco needed.

Another study from this same project using the same subjects found that children (at twenty-seven months) with a genetic predisposition to social anxiety (inferred from birth mothers' psychological history) did not show this problem if their adoptive mothers provided an emotionally and verbally responsive environment.[15] My responsiveness may partially explain why Marco's anxiety as an adult was moderate and not extreme. However, it's possible that my anxious personality contributed to his anxiety.

This research begins to explore how genes and environment interact, but it tells us nothing about the impact of other nurture issues (e.g., setting limits, consistency), nor does it compare the impact of parenting style to other forces that impinge on an adopted child, especially the external environment and, in Marco's case, early childhood sexual abuse. There is no data yet from the ongoing Early Growth study as to whether parenting has any impact on adolescent or adult outcomes, something that the established behavioral genetics research doubts.[16]

Contrary to Freudian theory, behavioral geneticists posit that the impact of the family fades by the time a child reaches adulthood. Adopted siblings as adults who share no genes with each other or with their adoptive parents are little more alike than two individuals picked randomly from the population, even though the siblings grew up in the same family.[17]

Robert Plomin in the sixth edition (2013) of his authoritative text *Behavioral Genetics* concludes that from childhood to adulthood the impact of the environment on adopted siblings in the same family decreases in importance from 25 percent to zero, while the genetic impact increases from 40 percent to 60 percent.[18] This means that genetics explains 60 percent of the difference between adopted siblings as adults who were from different birth families but raised together in the same adopted family. Harvard psychologist Steven Pinker concludes: "Whatever experiences siblings share by growing up in the same home makes little or no difference in the kind of people they turn out to be."[19]

If my parenting was not determinative in Marco's adult life, and if I couldn't prevent his troubles as an adult as I would have liked, I'm thankful that we enjoyed his childhood.[20] During our camping trips and traveling together in his preteens, I experienced the motherhood I'd dreamed about—sharing an activity and each other's company, while respecting each other's particular interests.

When Marco was ten, the two of us went up to Tuolumne Meadows in Yosemite and hiked to a High Sierra camp, where we pitched our own tent and paid for our meals and to take a shower. Having an extroverted son always led to interactions with other people, but we also enjoyed our own company and being together in solitude. We hiked to a lake where Marco fished. On the trail, he identified animal footprints and scat, demonstrating what he had learned in his Young Explorer program in Berkeley. He differentiated types of snakes and lizards, and he was excited when we saw marmots, deer, a fox, or a bear. I'd use my guide to identify wildflowers and Marco at least feigned interest.

When we reached a lake, sometimes after hiking five miles, Marco would unfold his fishing rod, take out the bait and hooks, and patiently wait for a bite, while I'd find a rock to lean against, with just the right amount of sun and shade, and pull out a book. Both of us basked in the vista before us—a pristine mountain lake with a snow-capped granite peak behind.

The next couple of years, Marco and I traveled together to the Southwest to hike the Grand Canyon and in Zion and Arches National Parks. I especially remember one experience when Marco was eleven and in the sixth grade. We had stopped for a couple of days in Sedona on our way to the Grand Canyon, before we would hike down to the Colorado River and stay overnight at Phantom Ranch.

In Sedona we hiked in the cozily warm, early April sun, through open fields with glistening grass and early wildflowers peeking out. Best of all were the glowing red rock buttes in the background, with hues from deep purple in the shade to auburn in the direct sun, set off by different colors of green trees—the dark of the evergreens to the chartreuse of the budding oaks. A few alabaster clouds wandered through the endless Western sky. We seemed dwarfed, but also embraced, by this almost postcard-perfect scene. When he asked, I couldn't explain why the sky in Arizona appeared so much vaster than in California. Even up in the Sierras, I never felt overwhelmed by the landscape as I did here.

So far, the trip had gone well. Marco was not yet old enough to be embarrassed by traveling with his mom, and not ready to express discomfort about being the only black person on the trail and in the restaurant, whatever he felt about it. Even in our camaraderie, however, race was just below the surface, separating us. When occasionally a person of color appeared—an African American, a dark-skinned Latino, or a Native American—he noticed. I was aware that he noticed, but I never said anything. Now I wonder why.

Marco often chatted with strangers, but I noticed that he never initiated conversations with African Americans we met while traveling. On rare occasions when an African American man crossed our path, Marco made up stories about him, imagining some actor from *The Cosby Show* or a famous baseball player. I never asked him about his reaction, maybe because I didn't want unpleasant realities to interfere with our good time. Probably, I just found it too hard to talk about race in any setting. On such trips, however, I got in touch with what it meant to be in the spotlight because of skin color. Strangers often came up and asked how I liked the current site compared to one we had visited the day before. I assumed they remembered me because I was a white woman with a black child.

On our last day in Sedona, I told Marco he could choose the activity. I knew he'd pick fishing, as he yearned to try out the new collapsible pole we'd bought for the trip. The motel suggested a nearby trout farm on an artificial lake, but Marco would have none of it. No novice, he wanted a real river. Our clerk said, "Nearby Oak Creek usually doesn't have fish, but with the spring runoff you might catch something."

Because of the steep walls of the canyon, the thick forest, and the cloud cover that day, we descended into gloom to the dark, muddy creek. As Marco bent to work preparing his fishing gear, I tried to make the best of the unpleasant setting and found a log to lean against to read my book. He easily caught several six-inch fish, but I'd never seen such ugly fish with huge mouths. "They're sucker fish, bottom feeders and inedible," Marco said and threw them back. When, after a couple of hours, he became discouraged, I suggested, "Why don't we go to the trout farm? At least you'll catch an edible fish, and we can take it to the Grand Canyon restaurant to cook."

We had done this in Tuolumne Meadows in Yosemite. Marco had been proud when the waiter ceremoniously delivered the fish to our communal table, where he presented a taste to all the other diners. I assumed all national park facilities would be similarly accommodating.

Marco reluctantly consented. When we got there, I had to agree that the commercial trout farm with its small, scruffy artificial lake was unappealing, as was the ease with which he immediately snagged two trout. He insisted on cleaning the fish himself, even though it was included in the admission fee. With the fresh fish in our Styrofoam cooler, we drove off in the emerging sunshine, happy to be on our way to the great canyon.

We arrived at Grand Canyon Village about 5:30 p.m., quickly unpacked, took a peek over the rim at the amazing site we would hike the next day, and hurried off to the restaurant, toting our fish in a plastic bag. As we entered, I asked the host about cooking the fish. "We don't do that here," he proclaimed. I pleaded with him, telling him about Yosemite, but he was unyielding. Tears rose in Marco's eyes. Bending down, I tried to comfort him. But he said, "If we don't eat them, I will have killed them for no reason."

The host, now moved but unbending with the rules, suggested that we buy some charcoal and go to the picnic area.

I was tired and my heart sank as I thought about the work in lighting a fire, but I said O.K. Finding the store closed, I had to come up with another idea.

"Let's go to the picnic spot and find someone with a fire. We can ask them if we can share their fire and we'll give them a fish."

A smile returned to Marco's face. By the time we found the picnic space, dark was descending, but many fires sparkled in the dusk. I was hungry, but leery about approaching strangers, something that doesn't come naturally to me.

"Come on," Marco said, "those people look nice." He pointed to a nearby glowing grill, with a family gathered around the table.

"I'd never do this without a kid," I said to myself.

I was relieved that they were accommodating, happy for us to use their fire, providing not only grilling utensils, but also leftover salad, potatoes, drinks, and dessert. As we chatted with them, their lantern sent out a warm hue in the dark, mild evening. Family members were too full to try our fish, but they agreed to put the leftovers in a cooler for their breakfast. Marco glowed. I didn't say I thought the fish tasted weird, nothing like trout fresh from the stream. Later, just before he fell asleep,

Marco seemed happy, but he murmured that the fish didn't taste as good as fish from a real lake. "I'll never again go to a fish farm," he vowed. I fell asleep quickly, glad that the crisis had been resolved and looking forward to the next day's hike.

When he was seventeen, Marco wrote, "Thinking back on my childhood, everything was sweet and easy. Mom was always there to treat me to all the good things life provided." Despite this reassurance, I still ask myself what I could have done differently to better prepare Marco and me for his difficult adolescence. I am consoled, however, by the knowledge that those happy times together cemented a bond that lasted despite the challenges that were to follow.

CHAPTER 7

Addiction

A s a teenager, Marco chose his friends and peer groups—at school and in extracurricular activities, in our neighborhood, and in the city. This was when our worlds began to diverge. Mine was a mixed-race neighborhood, close to the Oakland border, which facilitated Marco's exposure to people who looked like him. I was pleased when I learned that this was exactly the type of neighborhood that adoption professionals advocated for transracially adoptive families.

In looking at the friends he made in this neighborhood, at the public junior high school within walking distance, and the after-school activities in which he engaged, I will try to evaluate whether they met the criteria of behavioral geneticists who have found factors that can discourage or mitigate substance use in teenagers: participation in religious and after-school activities, a high degree of commitment to school, and peers who do not use drugs or alcohol.[1]

While still in the private elementary school, Marco began to play with the white, middle-class children who lived on our block. The girl next door was the daughter of two professors, and the boy and girl across the street had a stay-at-home mom and an engineer father. This couple began to ask the girl next door and Marco to accompany them to Sunday school at their Protestant church. The girl became a devoted Christian while Marco soon lost interest. Was this because of race? Would it have had a positive impact on Marco if I had found a church, especially a black church, that we could have attended together? In any case, Marco's minimal exposure to religion did not have a deterrent effect on his early teen substance abuse.

Through eighth grade, Marco participated in a number of extracurricular activities, but it turned out that they did not prevent substance abuse. During elementary school, Marco discovered another mixed-race boy being raised by a single white mom who lived on our street, a couple of blocks away. Marco met Matt when they were eight, not in the neighborhood or in school, but in the Young Explorer program run by the regional park district. They were the only two black kids. The YEs (as they called themselves) met most Saturdays during the school year for hikes in the East Bay hills and once a month they went on an overnight backpacking trip. The program culminated in the summer with a one-week trip to other parts of California. I liked the program because the male rangers, although white, were wonderful role models and the Asian American woman ranger modeled a competent woman. Half of the YEs were girls who demonstrated that they were as capable as the boys. Marco loved the program, and I saw it as a good environment.

Because I have always valued public education, I decided to switch Marco to public school starting in seventh grade, as many of his private-school classmates were doing. Marco could now walk to one of three public junior highs in Berkeley, one which was integrated by race and class. He often stopped to pick up Matt, whose house was on the way. They became closer friends. Well prepared by the New School, Marco was an honor student in seventh grade. Although he expressed little interest in school extracurricular activities, he played Little League soccer and baseball on weekends and joined the school wrestling team for its short season. His skinny body and light weight put Marco in a class where he usually won, as he built up his muscles. I saw this as a healthy response to his desire to avoid any kind of victimization. At the end of seventh grade, Marco told me that his greatest achievement that year was to have avoided a fight while still having others view him as capable of fighting. Would a private school have inculcated different values? Maybe, but Marco's addiction would probably have prevented him from living up to those values. I had made this choice, and Marco said he needed to learn to live in the real world.

By eighth grade, Marco had abandoned most extracurricular activities, thus shedding another protector from early-onset substance abuse.

But he still participated in Young Explorers on the weekends. Over the years, kids dropped out of Young Explorers and younger ones joined, but Marco and Matt stayed. They, along with three white girls and a couple of white boys, formed a YE pod. All lived in the Berkeley flats and not in the more affluent hills. Like sisters and brothers, they were never romantically involved. Although the program was supposed to serve children only from age eight to age twelve, the rangers permitted this group to stay, making them senior aides and evicting them only when they became high school students. I learned only much later that one of the white boys sold Marco and Matt marijuana. Thus, neither sports nor this wholesome outdoor organization protected Marco from easy access to drugs. Matt continued to be on sports teams throughout high school, but his participation did not protect him from drugs either.

A third factor that is supposed to protect youth from substance abuse is a high commitment to school achievement. Although he was intelligent and capable, by eighth grade Marco lost interest in school. He seldom skipped school in junior high and high school, for he loved to socialize and, as I learned later, that was where he sold marijuana and probably bought other drugs. I thought that Marco's dissatisfaction with high school was a question of finding the right fit. Now, I think he was on a quest to satisfy my strong commitment to education, while still looking for a peer group that would share his drug predilections.

School, rather than protecting him from drug addiction, became one of the factors that created a high-risk environment, one that made a youth like Marco with a genetic predisposition to addiction especially vulnerable. Such an environment for teenage substance abuse is created where there is easy access to drugs, to peers who use them, and where parents are unable to effectively monitor their young.[2] This task is more difficult for adoptive parents if they don't know the child's genetic background. Moreover, a genetic predisposition may also lead young people to be attracted to such an environment.[3]

Marco went to four different high schools, only the last of which worked to some extent. He had never wanted to go to Berkeley High, the only and very large public high school. It was racially integrated but internally segregated, and Marco believed he would have a hard time there. Knowing that the school had a positive reputation for academically inclined, high-achieving, mainly white students and a negative reputation for everyone else, especially black students, I agreed and still do.

I urged Marco to go to St. John's Catholic High School in Berkeley because it had a good academic reputation, and because the school was 50 percent black and drew students from all over the East Bay. From the beginning, Marco didn't like the school. Most of the black boys and girls were into competitive sports, which didn't appeal to Marco. The teachers with their conventional teaching methods bored him, and he didn't like their harsh discipline. Marco soon began to hang out with a group of white, alienated students. He told me later that they smoked a lot of marijuana. After discussion with a therapist, Marco agreed to stay through the end of the school year and then transfer.

I was surprised when Marco chose to transfer to University High, a small, private, academically oriented school in Berkeley with a mostly white student body. I later learned that Marco was attracted to the school's reputation as a countercultural enclave, which no doubt translated into tolerance for substance use. Marco lasted there only a year. The teachers were frustrated because he could do the work and get good grades, but he seldom applied himself. At the end of the year, the principal called us in and said Marco's lack of effort and poor grades meant they had to expel him. Marco could come back if he improved his grades at another school. The principal said nothing about drug use.

We made the decision jointly for Marco to go to Berkeley Alternative High School, but it was a bad one. There was not enough structure or organization. Students were expected to work on their own and meet with a teacher only a few times a week. Marco did enough school work to get credit for some classes, but he now had a great deal of time to hang out around Berkeley, with disastrous results. In the spring semester, he was arrested by an undercover policewoman in People's Park near the UC campus for selling marijuana. Since he was only seventeen and this was his first arrest, Marco was given probation with the proviso that he return to a more structured school. I was upset and disappointed, but I also felt guilty because I had approved of the alternative high school. I began to suspect that Marco's delinquency signified more than teenage rebellion and a reaction to adoption losses.

Marco's final high school, Oakland Urban School, was a small public school founded in 1973 through a grant from the National Urban League to permit small classes with a majority of teachers of color. While providing an academic curriculum, this highly structured school focused on personal development in a safe environment. Each student had a mentor-teacher and was placed in a small personal discussion group that met every day. Most of the students were black and from working- and lower-class homes. The teachers were wonderful, and the personal attention and support helped Marco. As one of only a few white parents, I was worried about how I would fit in, but everyone was respectful and responsive. Most important, this school recognized and sought to combat Marco's drug problem. After four months, Marco and I were called to a conference with the principal and some teachers.

I was impressed and grateful for their approach. The principal noted how intelligent Marco was and how influential with other students. But what they knew, and I hadn't known, was that every day he came to school high. He could avoid expulsion, they said, only if he went into an outpatient rehab program and stopped using marijuana before and during school. Marco agreed to go to a three-hour meeting four nights a week for two months, which I was also required to attend one night a week. Marco attended all the rehab sessions, stayed in school, and moderated, but did not stop, his marijuana use. His favorite teacher, a biracial woman, helped him prepare for the high school equivalency test, which he passed on the first try. Even though he didn't have enough high school credits, he was allowed to attend the graduation ceremony. A lot of his friends and I attended, and Marco beamed.

Having peers who use drugs is another environmental risk factor for early substance abuse. In all these different school environments, Marco always found those students who used drugs. This was especially easy in Berkeley and Oakland, but the 1990s was a time when the easy availability of drugs spread everywhere in the U.S. Behavioral genetics has shown that choice of friends and peer groups is not just the result of the school and community environment, Marco probably would have followed the same path wherever he lived or went to school. Behavioral genetics studies have found a large genetic influence on teens' choice of friends, whether these are preppy or countercultural peers, whether they use substances or abstain. The range of choices available to teens, as compared to younger children, permits greater expression of characteristics that are genetically influenced.[4]

After he left University High, Marco's main source of friends was not school, but our neighborhood and other community spaces in Berkeley and Oakland. I don't think he used drugs because of his friends; rather, he sought out friends who shared or tolerated his use. Behavioral genetics calls this *active* gene-environment correlation, where an individual, because of a genetic predisposition, seeks out an environment (e.g., friends) that shares his genetic bent, in this case, substance abuse. When others choose you for your genetic predisposition, it is called *evocative* gene-environment correlation.[5] Both active and evocative correlations characterized Marco's friendship network.

During his last two years of high school, Marco became involved with a group of Jamaican men who were about eight years older than he. I met two of them when Marco brought them home. One of them remained Marco's close friend into his thirties. I liked this caring man, who helped Marco and a number of other people, but he was involved in the marijuana trade. Despite my conflicted feelings, I often invited him over to our house for holiday meals. As his life became more stable and Marco's less so, I sought his advice about Marco and found his insights helpful. He said that the main reason Marco hadn't followed my advice or his to go to college was his use of drugs and alcohol.

Marco also became friends with a black brother and sister from a solid working-class family who lived around the corner. Through them he met Beatrix (Bea), a Filipino, two years younger, who lived in west Berkeley. She became his girlfriend for seven years, until Marco was twenty-four. I had often talked to Marco about my teenage years living in the Philippines and showed him pictures of me at age sixteen with my Filipino boyfriend. This may have had an influence on his choice, but having Bea as his girlfriend also avoided the black/white divisions in Berkeley.

Bea lived with her mother, an immigrant who had been a podiatrist in her home country and was now a doctor's assistant. Bea had a lot of conflict with her white stepfather and had spent some time in a juvenile hall. When Marco was just eighteen, and in his last six months at the Urban School, and Bea was not quite sixteen, she mainly lived at our house. After he graduated, they moved into my basement apartment for a year and then into her mother's small duplex. During these years after high school, Marco worked at various retail jobs and expressed no desire for a higher education. Aside from not charging rent when they lived in my house, I provided no financial support. I saw Marco and Bea regularly, had dinner with them once in a while, and included them in holiday meals, but I wasn't involved in their lives.

Bea was an ambitious young woman, who, like Marco, wasn't interested in school, but had entrepreneurial aspirations. I liked her and I tried to relate to her family, inviting them for a meal and to a holiday party. But Bea's mother made excuses about why they couldn't come and never invited me to her house. Living with Bea gave Marco some stability. He sold marijuana from her house, but he was never arrested during these years. Because she didn't use marijuana or drink, and Marco was often high or drunk, I found out later that Bea secretly skimmed off money from the marijuana sales. After they broke up, she bought a car, while Marco moved onto the couch of his Jamaican friend. Friends who tolerate drug use and sales can create an environment which is just as detrimental as having peers who use.

Even when I realized how deeply drugs affected Marco's life, I didn't blame these friends. Some, but not all of them, used marijuana; some, but not all, sold; and none was as addicted as Marco. I didn't discover until years later that Marco had a much wider circle of more casual friends who were involved with harder drugs, many from our neighborhood but most of whom I'd met only in passing or not at all.

Some behavioral genetics researchers posit that genetic influences on peer affiliation are greater than for most other behaviors.[6] It is not surprising, then, that researchers looking at adolescent substance use and abuse find that genetic predisposition is the most important factor in choosing a peer group of substance users, and that the peer group is more important than low parental monitoring or than school and community characteristics as a factor in addiction. Despite this conclusion, monitoring of the environment is something over which parents have more control. Effective parental monitoring can be a deterrent for adolescent substance use in a high-risk environment.

I tried to monitor Marco's after-school activities in seventh and eighth grade. In these years, Marco would hang out after school on our block with a group of boys, some from the neighborhood and a number of African American boys passing through. Marco's junior high served all of southwest Berkeley, and our street was on the route to the poorer, black neighborhoods to the west. If there was only a small group, I invited them in for a snack, but since our street had only small backyards or none at all, they then played ball in the street or went up to the park near Matt's house. I soon realized that Marco and I needed help in monitoring the after-school scene. Two days a week I didn't get home from work until six, and even when I was home, I couldn't keep track of the kids.

I hired Amalia, a nineteen-year-old Jamaican adopted by mixed-race parents, whom we had met a couple of years before, when Marco and I vacationed in Jamaica with friends who knew her parents. Amalia was now living with our friends in Berkeley and attending junior college. She wanted part-time work. I hired her to be at the house every day after school to keep an eye on things and make dinner on the days I worked. She could hang out in a way that I never could. Marco liked her, and I hoped that she could also help him with what experts said were the identity problems of interracially adopted teens. Perhaps she did, as Marco remembers her fondly, but I don't think she had much impact on Marco's life during junior high.

My parental monitoring had other limitations. I got to know only two boys from beyond our street, and I never visited their homes or met their parents. In retrospect, I realize that this is when Marco began to introduce me only to kids whom he knew I would like. For example, Marco became friends with a shy, African American boy, Billy, who along with his brother, was being raised by a single father. One day, after arranging with the father, I drove Marco to Billy's house in a poor neighborhood to play inside and stay for dinner. Afterward, Marco reported he was shocked that there was absolutely nothing in the house to eat until the father came home with food. Another day we took thirteen-year-old Billy out with us to get burritos. He had never had one, loved it, and saved half to take home to show his father and brother. Now, years later, I occasionally run into Billy, a reserved, decent man, working in various retail establishments. But the friendship between Billy and Marco didn't last.

Although it would have been hard as a working single mother, maybe even impossible, I wish I had gotten to know a few of these boys and their families. Even without having the same interests, we had our kids in common and I might have been in better touch with what Marco was doing. Also, he would have seen me interacting with a more diverse group of people than my white professional friends. I do know now that I underestimated how hard it was for a teenager to figure out where he fit into the complex race and class relationships of Berkeley and its surroundings.

When Marco was fifteen, in the first year of high school, he told me he would be staying overnight at a white friend's house in the affluent Berkeley hills. I didn't know this friend, and I reminded him of my rule that I had to speak to the parent first. He gave me a phone number. When I heard nothing by 8:00 p.m., I called the number and reached a mother who said that her son was staying at another friend's house for the night. Hearing my concerns, she said she would check. She called back to say that a big party was being held at a house where the parents were out of town. She called the police, who broke up the party, where alcohol and marijuana were being used, but they did not arrest any of the primarily white teenagers. Marco was angry that my phone call had initiated this action, but he didn't ask to stay out overnight again, and to my knowledge, he abandoned these friends. This is how I learned that teenage substance use was widespread in all classes and races in Berkeley.

Apart from this intervention, I was usually ineffective in monitoring Marco's exposure to drugs. Unlike many Berkeley parents, I never smoked marijuana, not out of moral revulsion but because as a nonsmoker it hurt my lungs. Although some of my friends smoked a joint now and then, and may have smoked when visiting the communal household, no one did so in my own house, the one I moved to when

Marco was four. I knew that Marco tried marijuana in junior high, but because of my naiveté about drugs and the permissive Berkeley culture, I wasn't concerned about it. I wasn't aware of how often or how much Marco used. When his grades began to fall drastically in eighth grade, I had him tested for learning disabilities (none) and hired a tutor, but I didn't attribute this change to marijuana. I certainly never suspected he was *selling*. This would have indicated Marco's greater involvement in a drug subculture, something I could not then imagine. Only in reminiscing at age thirty did Marco tell me that he and Matt began to sell marijuana in seventh grade.

"How did you have the money to buy it?" I asked.

"It was cheaper then and we saved up from our allowance."

'Where did you get it?"

"We bought it from Jim in Young Explorers."

"Why did you want to sell?" I asked.

"So I could buy the cool shoes that you refused to buy me."

When I'd noticed the shoes and asked where he got them, he lied, telling me he exchanged them for several shirts. Since I could not image him selling drugs, and since his clothes were always disappearing, I believed him.

I had caught Marco lying occasionally in middle childhood, but his lies (on many topics, not just about drugs) increased from age thirteen on. I had a hard time recognizing them. That Marco had an extroverted and open personality, and generally was not hostile to me, made it harder for me to spot the deceit. The most startling example was his secret marriage to Bea at age nineteen, which he revealed to me only five years later after she obtained her green card. Even my good friend who knew them well found it hard to believe that Marco had kept this secret. Maybe by then he was so used to hiding a big part of his life from me that he didn't realize I would have accepted his marriage and his reasons for it. Or maybe he gave in to Bea's and her mother's pressure to keep the marriage secret because they were using him.

I still ask myself why I wasn't more aware of, and more worried about, Marco's marijuana use in junior high. I think that the Berkeley culture of tolerance toward marijuana was part of it. I bought the view that marijuana is benign, even medicinal, and so I had little reason to condemn its use. But if I had done some research about marijuana and its impact on youth, I would have found out a number of alarming facts.

I would have discovered that marijuana in 1994 (when Marco was thirteen) was four times more potent than what I tried in the 1960s. By 2006, THC, the active ingredient in marijuana, was almost nine times stronger than in 1965, going from 1 percent THC to 8.8 percent.[7] I would have learned, too, that in the past kids were not exposed to marijuana at such a young age. By the mid-1960s only 5 percent of young adults aged eighteen through twenty-five had ever used.[8] By the turn of the twenty-first century, roughly 50 percent of young adults had tried marijuana and the average age of first cannabis use was fourteen. I would have been shocked if I had known that in eighth grade Marco was smoking several large joints of marijuana daily. According to the National Institute of Drug Abuse, in that year of 1994 less than 1 percent of those in eighth grade were smoking any weed at all.[9]

Research in the 1960s and 1970s documented that heavy marijuana use led to a motivational syndrome characterized by apathy, declining motivation, and decreasing ability to master new problems.[10] However, research in those same years reported that once one stopped using, these symptoms disappeared. This reinforced the common view in my circles that there was no long-term damage from marijuana. Most teens in Berkeley and elsewhere who tried marijuana did not become heavy users, and up to the 1990s marijuana was seen as nonaddictive.[11]

Experts now tell us that about 10 percent of marijuana users become addicted, mostly those who use it heavily and daily, compared to 32 percent of nicotine smokers, 23 percent of heroin users, 17 percent of cocaine users, and 15 percent of those who drink alcohol.[12] But for those who start marijuana use young, 17 percent become addicted, and for those who use daily, 25 to 50 percent are addicts, characterized by compulsive drug use and by long-lasting changes in the brain. A 2017 report that looked at the findings from more than ten thousand published studies on marijuana use found substantial evidence that starting use at an earlier age, being male, and smoking cigarettes are risk factors in progressing to problem marijuana use.[13] New research reported by the

Society for the Study of Addiction has found a genetic link to marijuana dependence.[14] I didn't know then about the heritage of addiction that Marco had from his birth parents, which made him more susceptible to a high-risk environment.[15]

We know now that marijuana may be especially harmful when started young, because the adolescent brain is still developing, particularly the prefrontal cortex, the site of cognitive impairments associated with heavy cannabis use. This part of the brain, responsible for impulse control, planning, decision making, and allocating attention, is not fully mature until well into one's twenties.[16] Thus, researchers conclude that a teenager who smokes marijuana daily may be functioning at a reduced intellectual level most of the time. Not surprisingly then, students who smoke marijuana tend to have lower grades, are more likely to drop out of high school, and more likely to be arrested for juvenile crime, all of which happened to Marco.

My failure to recognize Marco's drug use at an early age was particularly harmful since behavior genetics research has concluded that an environment with little parental monitoring allows adolescents greater opportunity to express genetic predispositions.[17] This research led me to question myself. If I had known more about drugs and had recognized and condemned Marco's marijuana use in junior high, would he be doing better now? After reading *Beautiful Boy: A Father's Journey through His Son's Addiction* by journalist David Sheff, my answer is "probably not," even if I had had a partner more knowledgeable about drug addiction.

Sheff himself used a lot of drugs as a young man, and although he never became an addict, he saw friends die of drug overdoses. As a result, he paid close attention to any early indications of marijuana or alcohol use by his young son and educated him about the dangers of drug use. David Sheff said: "Our close relationship made me feel certain that I would know if he were exposed to them. I naively believed that if Nic were tempted by them, he would tell me. I was wrong."[18]

Sheff was shocked when he found pot in his seventh-grade son's jacket and caught him lying about it. Sheff's being a white man with a white son and living in an upscale white community in Marin County didn't help him to block his son's early drug use. Even with more experience, money, resources, and family support than I had, Sheff couldn't stop Nic's later descent into full-blown addiction. Only in writing his book did Sheff recognize a probable genetic factor in Nic's addiction—his wife's father had died early of alcoholism.

I believe I made a lot of mistakes from the time Marco was age fourteen on. I felt out of control, which led to inconsistent parenting. I continually criticized Marco. After he was arrested at age seventeen for selling marijuana, I kicked him out of the house for several weeks, but then let him return and helped him get a sentence of probation only. I engaged in what behavioral geneticists call over-reactive (harsh, irritable, and angry) parenting, which is usually ineffective.[19] Behavioral geneticists' studies of misbehaving teens and adoptive parent response conclude that adolescents' inherited characteristics provoke negative parenting, which places the teen at further risk for deviant behavior.[20]

At the time, I heard about wilderness programs and schools for troubled teens, but I never considered sending Marco, because I thought an adopted youth would see it as another rejection. In recent years, in my adoption support group I have met two white couples—one with an African American son and the other with a mixed African and Native American son—and a single white woman with a mixed-race Filipino and white daughter, all of whom sent their acting-out adopted teenagers to wilderness and residential treatment programs. They all feared for their own and their sons' and daughter's safety at home because of violent outbursts, drug use, and trouble in school. The single mom and one of the couples felt the decision potentially saved their teenager's life. The other couple, on the other hand, felt it was a damaging experience leading to more problems between them and their son. Hearing these stories made me realize how individual these decisions are as well as their outcomes.

Periodically, I enticed Marco into therapy, but I didn't reach out to professional help for myself. Whether I could have been more effective at monitoring Marco or not, I certainly could have used more support. Why didn't I seek counseling for myself in these years? Like most social scientists, therapists, and parenting experts, I believed that nurture was everything. Parents, we believed, were the most important factor in how a child turned out. If a teenager was troubled, it was the parent's, especially the mother's, fault. Paradoxically, this idea immobilized me. I couldn't face the shame and guilt I'd feel if Marco's behavior indicated more than the temporary deviance expected of teenagers. I wanted to believe that the lovely boy I so enjoyed would return after this stormy adolescence. I went into denial mode, afraid that a therapist would blame me.

I now know that many parents (adoptive and biological) with troubled teenagers and young adults feel guilt and shame and are unable to talk about it. In exposing my own inadequacies as a parent, along with presenting a perspective that shows that genetic predisposition and environmental factors are as important as or more important than parental input, I hope to enable other adoptive parents to seek help sooner than I did. Many years later, when I joined a support group for parents of young, overwhelming male adults who had failed to launch, I realized how earlier I would have benefited from being in a group of parents facing similar problems, and just how alone and overwhelmed I had been. If adopted or biological parents become aware of what behavioral geneticists have discovered, they can without guilt accept help on how to better monitor their teenager's environment so as to lessen the risk for detrimental genetic expression.

One environmental factor that I have not seen addressed in either behavioral genetics or addiction literature is how violence in the drug subculture affects young users. Certainly, it contributes to a high-risk environment. My experience makes me believe that genetic factors had a large impact on how Marco avoided becoming the victim of violence. But my stable household after Marco was four could have helped. A Princeton University study found that family instability increases aggressive behavior, especially in boys.[21] Luck, no doubt, was another factor.[22]

During the last decades of the twentieth century, violence increased in all sectors of society, but the change was most severe in inner-city black neighborhoods. Violence increased disproportionately there because of the rapid decline of industrial jobs and because of the war on drugs. In *The New Jim Crow*, legal scholar Michelle Alexander makes a convincing case that this war on drugs, which began before the spread of crack cocaine, focused on arresting inner-city black men for nonviolent drug possession. The result was a huge increase in the incarceration of poor, African American men and the decimation of family and community ties in black communities. A prison record isolated black men, who, unemployable and desperate, often turned to violence.[23]

Even though our neighborhood is not "an inner-city black neighborhood," these social forces affected Marco from an early age. My upbringing in a small, overwhelmingly white university town in upstate New York in the 1940s and 1950s didn't prepare me to raise an African American boy in an urban environment in the 1980s and 1990s.

In the mid-1980s, when Marco was four years old, we came home one day to see police cars and a huge commotion a half block away in our mixed-race and class-diverse neighborhood, one which I considered relatively safe. When we walked over, we were shocked to see a parked car with a dead black man falling out of the passenger's door. The police told us that the shooting (probably drug-related) had taken place in a nearby neighborhood and the car and body had been dropped there. I had never seen anything like this as a child or as an adult. I was horrified and didn't think to talk to Marco about what he felt. What did it mean to him that the dead man was black? How did he make sense of it? I never tried to find out.

Six months later, a friend and I were on an East Bay beach, playing with our young sons, when a dead African American man washed up on shore. We grabbed the kids and ran to a nearby house to call the police. Though we tried to shield them, they were fascinated by the body. We learned later that the dead man was a victim of competition among drug dealers.

Yet, I thought of this violence as anomalous, and in my white, middle-class world, it was. I still believed that life within the home and personal relationships were more important in influencing children than the external environment. Because of my background and beliefs, I didn't think about how a black boy might find a different meaning in such experiences. I now believe that seeing dead black men might have led a small boy, without a father active in his life, to feel more vulnerable and to believe that he would have to protect himself, probably affecting Marco more than my friend's white son. These events seemed to reinforce Marco's attraction to the gun culture he saw in the media and at the toy stores. Even though I forbade violent T.V. and movies, and forced Marco to throw away the weapons of his Star Wars characters, my strong commitment to nonviolence faced a huge disadvantage. My message was in conflict with a culture, especially a male culture, saturated with violence.

The years from 1992 to 1994, when Marco was twelve to fourteen, saw the greatest number of homicides in the U.S. of, and by, both black and white males under age twenty-five. A 1993 national survey by Harvard University of children in grades six through twelve found that 57 percent of young people said they could get a gun if they wanted one, 39 percent knew someone who had been killed or wounded by gunfire, 35 percent believed that they were likely to die by a gunshot, 15 percent had carried a handgun in the last thirty days, and 11 percent had been shot at during the past year.[24]

I did not know these statistics at the time, and if I had, I would have found them hard to grasp. Even now, I find it difficult to comprehend that guns are so available to young boys. When Marco was thirty, I asked him when he had first owned a gun, knowing from earlier revelations that he owned one at age nineteen when he was selling cocaine.

"I bought one in eighth grade and hid it under my bed," Marco replied.

I was taken aback. When I told this to a friend, she asked, "Didn't you regularly inspect his room?" No, I hadn't, and even in retrospect, I wouldn't have. A parent is not a cop, and I believe that children by age thirteen have a right to privacy in their own room.

"Why did you want a gun?" I asked.

"I never used it, but I took it to school sometimes, because I wanted to be cool," he said. "I was also scared." I was sad at this revelation. Would he have felt safer with a man in the household?

Marco sold the gun before the school year was out, and owning a gun has not been a major part of his life since then. Maybe my proselytizing had some effect, or maybe he figured out for himself what expert studies prove: gun owners are more likely to get shot than those who don't own a gun.

More likely, he has a genetic temperament that is not aggressive or prone to violence, a temperament like that of his birth father. Marco verbalized an attraction to violence in order to be cool, but he did not practice it. Once, when he was fourteen, I heard him on the phone bragging about a fight he was in on Telegraph Avenue. When I questioned him about it, he showed me his bruise-free body, telling me that he stayed a block away. Marco's pattern of boasting about violence, while trying to avoid it in practice is still true of him today.

Only once did Marco act violently in my presence, and that was after he started using cocaine and heroin in his early twenties. Once, when he was high on heroin, he broke the glass window in my front door when I would not let him in, and then ran away. If anything, having a man in the house, depending on what kind of man, might have made Marco act more violently, if he felt he had to prove his masculinity. Over the years, I saw more evidence of a self-protective stance in regard to violence.

He never confronted police, even when he was stopped for no reason other than being a young black man, and even in the most humiliating situations. When Marco was about twenty and still with Bea, they got on a bus in downtown Berkeley to come to my house to have dinner with Amalia, who was in town for a short visit. Halfway down Shattuck Avenue, police with sirens stopped the bus and ordered Marco and Bea off. Fortunately, they had no drugs on them. The police had them sit on the curb on the busy street for half an hour and allowed no phone calls while they checked them out. Finally, they let them go, saying only that Marco resembled a man who had staged a robbery an hour before.

Even when he was high or drunk, Marco was polite to police. One night in his mid-twenties, Marco was found at 3:00 a.m. in downtown Oakland by two Oakland policewomen. He was wandering around drunk, with no shirt on. He had no idea what happened to his favorite shirt, the red one with the embroidered elephant I had brought him from India. He gave the police my phone number, even though he wasn't living with me, and they called.

"If you come and get him," the policewoman prompted, "we won't have to charge him or put him in jail." But I was finished with rescuing him.

"It's too late," I objected, "and I'm old. Why doesn't he call his friend?"

"Well, if we bring him to your house, can he sleep there?" she begged.

I reluctantly agreed. Thirty minutes later, the blond and cheerful policewoman led him, shivering and shameful, up my front steps. As I stood in the doorway in my tattered blue bathrobe, arms crossed, looking grim, she remarked to me, "Don't be too hard on him; he was very polite to us."

When Marco became more heavily involved in the drug subculture, he still used caution. After less than a year of selling cocaine when he was nineteen, he quit because he saw its dangers. He and Bea were still living in my basement apartment at that time, and they were robbed at gunpoint by someone who knew Marco was selling and would have a lot of money around. I was on a trip abroad and should have been suspicious when I came home and found they had adopted a pit bull; however, at the time I didn't know that drug dealers often kept such animals for protection. All I knew was that this dog scared me, my tenant in the backyard cottage, and the neighbors. My demand that they get rid of the dog was what prompted them to move to Bea's mother's house. She let them keep the pit bull. Marco sold only weed there, nothing heavier. Unlike friends who grew marijuana, Marco never did, although he often worked for wages for a small grower during harvest seasons once growing medical marijuana became legal in California.

Once when Marco was in his late twenties, I tried to be positive and keep communication open by remarking to him, "You've achieved something important. For ten years you've used, sold, and worked only in the underground economy. You've spent only a few days in jail, and never been arrested for a felony. You've never been hospitalized, taken an overdose, or been wounded."

He bragged in response, "You don't know how many times I've had a gun pointed at me." While there may have been male exaggeration in this statement, there was also truth. Hence, luck, not just his temperament, accounts for his survival in his risky environment. His friend Matt, the biracial boy who lived on our street and was his friend since Young Explorers, was not so lucky.

At age thirty, Matt was gunned down in his backyard in a middle-class neighborhood in central Berkeley, in the middle of the day. Matt was partnered, had two sons, and had almost finished his electrician's apprenticeship, but he grew marijuana in his backyard and had been selling it for years. The police have never solved the murder, but they have stated that it was not random. Matt's was the only homicide in Berkeley that year, but the adjacent city of Oakland recorded 103 murders, most of them young black men killed with firearms.

According to Marco, Matt was never involved in using or selling hard drugs as Marco had been in his late teens and early twenties. Because of his settled lifestyle, Matt always appeared to me to be in less danger than Marco. Matt's marijuana growing, however, made him vulnerable. Even a large patch in one's backyard in Berkeley seldom resulted in police intervention, but growing illegal marijuana is a dangerous enterprise, for it's worth a lot of money and is easy to steal. When harvested, it could bring in five to ten thousand dollars at a time for a small operator like Matt. If you are reluctant to put your money in a bank, fearing questions about its origins, and if you are open, friendly, and trusting, you expose yourself to the greed of others. I don't know if Matt owned a gun, but the farming of an illegal substance, with profits unprotected by the law or the courts, encourages violence, especially in a society where guns are so available.[25] I asked Marco if Matt would still be alive if he hadn't been growing marijuana. "Probably," he replied.

CHAPTER 8

Reconciliation

I n the years after Marco returned from living with his birth parents in Louisiana, addiction ruled his life and I struggled with how to respond. For over nine years, I went through the ups and downs common to those in families with addictive members. I relate here some incidents that illustrate the roller coaster of my fears and hopes.

A few months after coming back to Berkeley when he was twenty-eight, Marco gave in to my pressure that he go into rehab and entered a Berkeley outpatient program with a good reputation. It was a year-long commitment, but he completed only five months clean and sober, going to the program every weekday, living in a clean and sober house, and attending AA meetings. He said he went into the program for me. He probably would have been more successful if he had wanted it for himself.

In the first couple of months, he took rehab very seriously. He kept a journal, which I discovered once when cleaning out the basement and which he gave me permission to read. He wrote that he felt wonderful physically when he was off drugs and alcohol and psychologically he was better too. His anxiety in new situations declined, and his paranoia that everyone was looking at him decreased. But he expressed a lot of anger, directed especially at other men in the program and in the house. Was he indirectly working out his anger and disappointment with his father? Was this anger a residue from the child sexual abuse? He had some therapy as part of the program, but probably not enough. He refused to go to a private therapist.

The rehab program was in the morning five days a week, He also attended a few AA meetings in the evening, along with house meetings. But he had too much time on his hands.

He stopped seeing old friends because they all used drugs and alcohol, but he didn't make new ones. He had started a relationship with a new girlfriend before he entered rehab and now he saw her more and more. He began to write in his journal about days when he was up and others when he was depressed, which seemed to correlate with the ups and downs in his relationship.

After a couple of months, Marco became bored with the material for study in the program and didn't like the rigidity of the routine. He relapsed, the first time confessing to his counselor and getting praise for telling the truth. But later he started missing meetings, probably to try to recover from his use so he would not have a dirty piss test. He complained to me about how much he hated the clean and sober house and the poor neighborhood in which it was located. He didn't like his roommate, especially when they were moved into a smaller house. He wrote: "Today, there was another dead body out by the house. I don't want no part of this west Oakland funk."

I let him move back into my house, a short bus ride from the program in downtown Berkeley. My motivation was partly selfish, as I needed someone to take care of the dog and the house while I went on a two-week vacation. While I was gone, he carried out his responsibilities for house and pet care, but didn't take care of himself. This is when he started missing program meetings and spending more time with his girlfriend.

After five months he dropped (or was forced) out of rehab and slipped back into addiction. Where once he had sold hard drugs but didn't use them, now he didn't sell but used cocaine, heroin, meth, and other drugs, and always cigarettes, alcohol, and marijuana.

Reconciliation, for me, meant trying to enjoy meaningful experiences, however transitory, with Marco. The joy and pain in our ongoing relationship are illustrated by our joint experience of Matt's memorial service and our continuing grief.

Two girls, friends from their years together in the Young Explorers, organized the first informal memorial for Matt in Tilden Park in the hills above Berkeley on a Saturday four days after Matt's death. All three rangers who had guided Matt and Marco in Young Explorers attended. When others in Matt's black family and mixed-race communities heard about it, they came too. Even those who might not have understood Matt's interest in outdoor activities showed up.

The March day was sunny, although still cool. As I drove Marco and Crystal, his new girlfriend, to the park, I saw that he was unshaven, his long dreadlocks were unstyled, and he wore a dirty black peacoat over his ragged jeans and sweatshirt. Since he usually dressed well, I presumed his appearance indicated he was using. I was uneasy about what others would think of him and of me. I tried but was unable to control myself from commenting in the car about how unkempt he looked. I said that it was disrespectful to dress like that. He said nothing.

When we arrived, we entered the crowd of about seventy-five people near the picnic tables loaded with food and clustered at one end of a sparkling green field, near a grove of eucalyptus trees. Surveying the crowd, I noticed that most of the mourners stood in clusters or sat in groups self-segregated by race and social class, although some people flowed from one group to another. These groupings represented the achievements and limitations of integration in Berkeley. Perhaps I made these observations because I was a sociologist. More likely, it was a way to distract me from my uneasiness about Marco.

One cluster consisted of a mixed-race group of young people in their late twenties and early thirties, sitting with Matt's white sister and her Filipino husband, along with black, white, and other friends of indeterminate racial heritage. Another distinct group consisted of black working- and middle-class extended families, one organized around Matt's high school girlfriend, the mother of his ten-year-old son, with her parents, siblings, cousins, and their families. Matt's current partner attended briefly with their seven-year-old son, her two teenage daughters, and other family members.

I made an effort to talk to some individuals in many of these groups, especially Matt's partners and his sons, but I was an outsider. My place was in the group of middle-class whites, old and young, which included Matt's mother Susan and her friends, the rangers, some of Matt's former teachers, parents of YEs, and former YEs themselves now in their early thirties, some with partners and small children.

Not all whites, however, were middle-class. I was surprised to see Jane, an ageless woman with mental health problems living on SSI, who used to rent a flat on my block. Everyone in the neighborhood knew Jane because she spent a lot of time walking her dogs, befriending the neighborhood dogs, and talking to just about everyone. She was here because her son James was a friend of Matt's. I was startled to see Larry, a man in his fifties, white and poor, and in worse physical shape than when I had met him eight years ago. Marco had introduced Matt to Larry, an independent car mechanic, who was the stepfather of Bea, Marco's former girlfriend. Larry and Matt bonded over their love of automotive repair. When Larry became homeless after a divorce and a stroke, Matt let him and his dogs live in a van outside his home, until the neighbors complained to the city. Now, Larry spoke publicly in tears about how kind Matt had been to him, often bringing a plate of food to the van.

The group that held most interest for me was Matt's black male friends. I knew they were important to Marco, but I recognized only a couple. Ten men, many light-skinned, maybe half with white mothers, several of whom grew up in our neighborhood, all of whom Marco knew, stood together. Most were better dressed than the more casually attired whites. These men wore black slacks with dark shirts or jackets. Most had short hair, while a few sported cornrows neatly braided or short dreadlocks. To my embarrassment, Marco looked scruffier than anyone else, but he redeemed himself by being as outgoing as ever. He spoke with the rangers, and he hugged and greeted some of the other YEs and their parents, but he gravitated to this group of black men.

It struck me that I barely recognized most of them and knew little about them, even though many had lived nearby as boys. I knew less about them than I knew about most of the whites gathered there. Later, I asked Marco to fill me in. He pointed out one who owned his own plumbing business and another who was heavily involved in illegal activities. Most, however, like Marco, worked in the underground, quasi-legal economy. They collected scrap metal, worked as handymen and gardeners, and repaired cars or other appliances, all off the books. Many, in addition, grew and/or sold marijuana. Even though I could understand why black men might end up in this kind of work while white men did not,

I couldn't then—though I have since—let go of my middle-class values about the benefits of salaried employment.

Along with many of these men, Marco got up to speak. My shame about his dress and addiction vanished as with great eloquence and grief he spoke about his friendship with Matt. "He was my guardian angel," Marco said as he wept and gestured to the sky.

Whenever I was in trouble, I'd call and Matt would come pick me up and try to straighten me out. Reminding me of my talents, he urged me to go to college, even though it wasn't for him. I never got into trouble when I was with him, and we had so much fun. One day last summer we took his son fishing and stopped by the Acme Bakery to stuff ourselves with fresh bread and rolls. We talked about going backpacking again. Matt said to me: "We've spent so much time becoming black that we've forgotten our white roots. We have to get back to them."

Marco concluded, "Now I'm lost."

I was pleased that Marco spoke publicly about being half-white, which he enlarged on when he spoke again: "I was drawn to Matt because we both had white moms—although I was adopted—and in our day, light-skinned black men were not cool. I'm happy that my mom is here today. Where is she? There she is," he said, pointing me out.

Whereas I had been embarrassed about how he looked, I was now proud to be identified as the mother of a son so open and articulate, happy that my black son could publicly acknowledge his white mother and his adoption. I forgot then that this articulate, compassionate young man was not often in view. His mood and interactions varied by how much alcohol, or what drugs and how many, he consumed on any given day.

Several months later, on Mother's Day, Marco and I went for a hike, something we often do on this day. Despite his addiction, love of the outdoors brings us together. On this cool, sunny Sunday, Mitchell Canyon, on the north side of Mt. Diablo (the highest peak in the Bay Area), was topped by rolling emerald hills and filled with white mariposa lilies with blood-red patterns inside and fragile, pale yellow globe tulips that grow nowhere else in the world. Marco told me he had brought along clam shells to make a memorial in the woods for Matt. I was touched, for it showed his love for Matt. As we walked, he picked up and added unusual stones and pieces of wood.

"Why shells?" I asked.

"Because Matt loved the beach," Marco replied. "He also liked hiking in the woods," he continued, "and the shells will stand out here and not disintegrate so fast."

On the return trail not far from the entrance, Marco suddenly turned around and spied an animal on the path not far behind us. "A coyote," he shouted. "I've never seen one before." I was sure that the coyote would disappear, scared off by the noise, but it stood there and then slowly ambled off the path. A few minutes later, we spotted the coyote in the grass, prowling slowly along, parallel to us. "That's Matt," Marco said. "His soul is in the coyote and he's thanking us for remembering him."

Although I'm a nonbeliever, Marco's connection to his dead friend reinforced for me his ability to keep connections to me and many others, despite all the disconnections in his life history. This was a beautiful moment. I was proud of him and immensely touched.

But such moments were rare, as I struggled with how to keep a connection to Marco, a connection that was neither codependent nor enabling, while going on with my own life. I made a clear decision that he could not live in my house. His addiction was too upsetting and depressing for me to live with in my space, and I had given up the idea that I could have any impact on his using. But I had trouble deciding when to help him financially, when to pay his rent or buy his groceries, and how to hold the line against his constant pleas for small amounts of money.[1]

I had hoped that Matt's death would be enough of a shock to Marco that he would decide to turn his life around. Toward that end, for a year afterward, I paid for a studio apartment and bought groceries. Having these basic necessities taken care of would, I hoped, facilitate his return to rehab, to finding a job or going back to school. Instead, his grief led him deeper into addiction. He moved his new, African American girlfriend, Crystal, into the small apartment. He later admitted that he was attracted to her because she was also an addict. I soon found out she was a prostitute and had a personality disorder leading to hostile outbursts and paranoia. I was appalled. I'm sure Marco picked up my feelings, even though I didn't say anything. I knew that trying to influence his relationship would be counterproductive.

Her behavior, coupled with their drug use, led the landlady to evict them. I paid for another apartment in a larger building in Oakland, but the leasing company soon broke the lease, returning some money to me and forcing them out. What followed was at least six months of homelessness, followed by a year or more when the two of them lived in cheap motels in down-and-out areas of Oakland.

During this period, I cut off contact with Marco for three months, unable emotionally to handle his life of addiction. I was angry and sad. But I loved him and couldn't stand losing contact with him, so I started meeting him occasionally. One time, after lunch, he asked me to come to see the homeless shelter he had built. The wooden structure covered with tarps and furnished inside, he told me, with a platform bed, a couch, rugs, and other amenities made him proud. It was on a grassy area near the freeway. I looked at it from a distance, out the car window. I praised his creativity, but I couldn't bring myself to walk over and look inside. I never imagined that my son would live like this. A few weeks later, the shelter was bulldozed by Oakland city crews ordered to get rid of homeless encampments. This made me angry and sad, too, because the city provided few units of alternative housing. Soon Crystal and he moved into cheap motel rooms, paid for by money Crystal earned and what he made selling drugs and used goods he scrounged off the street. I never went inside them either. Just the thought of what his life must be like was too painful for me.

Four years after Matt's death, Marco received a DUI at 4:00 a.m. for driving intoxicated and without a license. I was in a hospital, thirty miles away, with a broken hip. I had slipped on gravel and fallen on Mt. Diablo on another Mother's Day hike with Marco. Although his license had been suspended again (for reasons that I no longer remember), I knew he was not high, had not been drinking, and was a competent driver. So I asked him to hike back and drive the car to the nearby hospital while I was flown in by helicopter. He was helpful and concerned, but I wanted him to go home. I was in no pain, and my friends would be with me at the hospital the next day, before and after surgery. Assuming that he would stay sober, I asked Marco to drive my car back to Berkeley, park the car in my driveway, and give the keys to my tenant. Instead, I learned later, he did some drinking and took my car to the motel.

After surgery the next day, I got a phone call that he was in jail and my car had been confiscated. I refused to bail him out, and my friends and I went through the complicated bureaucratic process to get the car back—document signing from the hospital, giving friends written permission to get the car out of police custody, credit-card signing, and key retrieval from my house. I was angry and depressed. How could Marco, who had been scared but emotionally supportive of me after my fall, have done this?

I tried to reconcile myself to the Alanon slogan, "I didn't cause it and I can't cure it." I stopped giving Marco money on a regular basis and stopped pushing him to go to rehab or do anything else. I tried to keep my distance and to do a few acts of kindness as a gesture of reconciliation, especially around Thanksgiving and Christmas.

Holidays were always important to me. When I was a single woman in my mid-thirties, I sought an alternative to flying east to spend the holidays with my parents and siblings and their children. I became a prime promoter of holiday dinners with a network of friends—coupled and single—who did not have family nearby. I organized potluck, sit-down dinners, or buffets at my house on one or both holidays, sometimes with up to twenty-five guests, and encouraged other friends to host too.

I had continued this tradition after I became a mother. Because of Marco, I now wanted to include some people of color, but my friends and colleagues of color had their own family traditions, so most holidays at my house or my friends' remained white.[2]

When Marco was young, some people brought children, but later, the regulars became singles and couples without children. Not only were they all white, but they were mainly academics. I remember one Christmas when we were a small group of eight adults and Marco. He asked: "Did everyone here write a book?" One of the husbands said no, and could tell Marco about his work as carpenter-contractor. But I noticed, as I looked around, that he was the exception. His wife, who was an academic, had published a book, as had everyone else.

In high school, Marco often preferred Thanksgiving dinner at the extended family of my old friend Katrina and her son, Kevin, whom we had known since Marco's early childhood. One of her nephews often babysat with both boys when they were small, so this extended family seemed more comfortable to Marco than my academic friends. Later, in his late teens and early twenties, I encouraged Marco to invite his friends to our house for holiday dinners. I recall an especially memorable Christmas dinner when he invited three male friends (one white, one black, and one Latino) and their girlfriends. They had their own table, enjoyed themselves, and left early to go to other parties, which was fine with me. But Marco lost most of these friends because of his addiction, so this never became a tradition.

As Marco's addiction deepened in his late twenties, I was relieved when he had other plans for the holidays. One Christmas day he turned up drunk and high for a small dinner I was hosting. He was unpleasant, ate quickly, and walked out with angry words. The next year, I told him he was not welcome. I would not let him ruin my holiday and embarrass me in front of my friends. The next Christmas, I brought a holiday dinner I had cooked to the motel where he and Crystal were staying before I went to dinner at a friend's house. I brought a ham, sweet potatoes, beans, salad, rolls, and dessert, along with a bottle of sparkling apple juice.

Two years later, Marco and Crystal moved in with an elderly man for whom Crystal worked. His modest house was only a couple of blocks from mine. From there, Marco often dropped by to do household chores for me, using his handyman's skills. After a year, Crystal had mainly moved out, but Marco continued to live with and help care for the man, who was in his early nineties. When Marco was thirty-five (in 2016), he and Crystal officially broke up. Since then, he has declined to have a new romantic connection even when women pursue him. "Part of my problem," he said, "is that I've always been in a relationship and never spent time working on myself."

After this, Marco, for the first time in many years, came to both Thanksgiving dinner and my Christmas Eve party that year. On the latter occasion, Marco dressed nicely and interacted with friends of mine, many of whom he hadn't seen in fifteen or twenty years and others he had never met. He was anxious, but his excellent social skills kicked in. As important, he spent most of both days helping me set up for the parties and then did a lot of the clean-up afterward. Both his help and his participation in the parties made me happy. But during the holiday season, he also went on a two-day bender.

Marco seemed happier when he was employed and had money, but he had not had a job in the established economy for more than ten years. Instead, for several years, he had regular employment on marijuana farms, where he was pleased at his success in learning farming and carpentry skills (e.g., building greenhouses). I thought that once the farming and selling of marijuana became legal in California, perhaps the skills he'd learned would lead to some work stability, even though it was not the best field for someone with a history of addiction. I did not take into account how large corporations and the mafia would move in and undercut small growers.

In November of 2017, Marco came home for Thanksgiving after the fall harvest, saying he would be returning to the farm near Lake Tahoe to finish cleaning marijuana and preparing for the next planting. But then he kept giving excuses about why his departure was postponed. In the meantime, in December, a death in his birth family led to a January memorial service, delaying his departure again. But then he never left. I learned that he had had a falling out with both his employers, who had become friends, and who Marco claimed owed him thousands of dollars. This reinforced my negative assessment of work in the drug world, even though growing marijuana had become legal.

Why didn't I enforce the rule that he couldn't live in my house? It was winter and the elderly man he had lived with had moved into a nursing home, so his house was no longer available. Marco seemed isolated and depressed, so I let him stay on. A few times he moved in with different friends (some black, some white) in the neighborhood, and I paid a little rent. But these arrangements never lasted.

His presence in the house made me more and more depressed. He slept late, got up and hung out for a while, did some household chores, then left in the afternoon, probably to scavenge and sell what he found or to hang out with his drug dealer friends near the Ashby BART station. He came home for dinner. I kept food in the house, didn't cook for him, but sometimes gave him leftovers. Then he went out at 10 p.m. and came home about 2 a.m. I knew that many nights he hung out in front of a neighborhood bar selling marijuana. Or at least that's what he said he was selling. I knew he was drinking, but only later did I discover he was still using heroin and cocaine.

Over the years I had lots of help (friends, group and individual therapy, and Alanon meetings) in trying to figure out how to deal with Marco's addiction and my roller-coaster of feelings about him, but I now had the best help I'd ever had. Two years earlier I had joined a support group for parents whose adopted sons and daughters are troubled adults—a group sponsored by an Oakland adoption agency that offers parent education and support. The eight adoptive mothers in this group (including the facilitator) and the one heterosexual couple have adult children from age twenty to age forty-five, who are white, black, Asian, and mixed race (white with African American or Native American or Filipino heritage). We are single, divorced, widowed, and coupled. What makes this group so effective is that we all face similar dilemmas in how we relate to our adult sons and daughters. We all have found that setting limits is important for them and for us, but we reject a simplistic notion of "tough love." We all are struggling with the issue of how to maintain contact and support, without letting their troubled lives disrupt our own or deplete our resources. Meeting with this group every week and receiving their empathic feedback, I was able to see how stuck I was and how ineffective.

I decided I needed to go back to see a therapist. After a few sessions, I decided to try one more time to get Marco into a rehab program, this time one that was inpatient and expensive. The therapist didn't have an opinion about whether it was a good idea or not, but she said that if I was going to try, I had to give Marco a firm ultimatum that he had to move out and that I would no longer provide financial support if he turned down rehab. I also had to recognize that the expense—paid for out of my home equity—might be for naught.

In adoption theory and practice, adoptive parents are portrayed as privileged because they have more economic resources and higher class status than birth parents. This was true of me, although my resources and status were based on education and on the luck of being born during World War II into the baby-bust generation.[3] Adoptive parents are not privileged in terms of their tribulations in raising children.

Since Marco liked the country and I wanted him far away from the current environment, I looked for a rural rehab within driving distance. With the recommendation of a psychiatrist who had seen Marco, I found a place in Sonoma County that seemed to fit him. The modest accommodations of Olympia House, in a beautiful but isolated setting, had a flexible program with lots of individual therapy. The program created a continuing community, inviting alumni back for Thursday-night dinners and a group discussion, for an AA meeting once a week, and for tune-ups two or three times a year. For the latter, alumni could come back for two nights and three days to participate in the program for free.

I presented an ultimatum. Marco grumbled but said he would go, although he needed a week to get ready. I acquiesced but set a firm date. Fearing he would back out, I asked his oldest and best friend, a Jamaican American, Paul, if he could take off work on a Monday morning to drive up with us. Paul, who had seen the negative impact of Marco's addiction, wanted him to go and agreed to drive my car. I was relieved, especially since over the weekend Marco came home with bruises on his face, saying another man had attacked him. Marco was drunk and probably high.

When Paul arrived, Marco was still packing. Nevertheless, we managed to get him to intake, while Paul and I were given a tour of the farm. We left Marco sullen and uncommunicative. As we drove away from the very white environment, Paul said to me, "I'd say there is a 50–50 chance that Marco will walk away from here, as he's a good hiker." I agreed.

We were wrong. Before the week was out, Marco called to say that he loved it there. When I visited after the second week, both staff and other patients reported that he was the most enthusiastic participant, attending all the groups, doing all the homework, open in therapy, attending meditation and nondenominational Sunday services, and going on the beach outings. He loved the good food and began working out daily in the gym, gaining weight, which he needed. He stayed five, rather than four, weeks and would have stayed longer if I could have afforded it. Then he entered their outpatient program for another two months. While in the inpatient program, Marco made friends and entered a sober living facility in Marin County with one of them. This friend had a car and they drove together back to the outpatient program. With the other eighteen residents of the sober living house, Marco attended AA meetings at least six evenings a week.

Now, Marco has been clean and sober more than six months with no relapses. He seems committed to stay that way and change his life. He has a full-time job at Whole Foods Market in San Rafael—the first time in twenty years he has worked in the aboveground economy. He is figuring out how to make a transition to living on his own or with a friend or roommate while continuing to be a part of the AA community.

Looking back at some notes I wrote when he was in the outpatient rehab in Berkeley eight years earlier, I noticed that he never made a friend there or in the sober living house where he lived in Oakland, despite many more African American participants. He has also matured, and even though he got into a fight before going to rehab, he is much less angry and hostile than when he was younger. Perhaps the creative and more individualized program at Olympia House engaged him as the Berkeley outpatient one did not. Probably all these factors, along with the emphasis on community, have contributed to his ability to sustain sobriety.

Thanksgiving 2018 included Marco's friend Paul, his wife, and their six-year-old son, along with my surrogate daughter, Sasha, her husband, and their twin boys, who were five and a half. I was the only one of my generation, but the smaller numbers (six adults), the racial integration, and the well-behaved but lively children made it special. I was pleased that Marco's world was more racially integrated than mine. This year, Marco helped with the cooking as well as setup and cleanup. On Friday morning after Thanksgiving, he asked me, "How does it feel to leave a half bottle of opened wine in the refrigerator and come down the next morning and find it still there?" "It feels great," I replied.

Alanon tells us that we didn't cause it and we can't cure it. This is certainly partially true, but I couldn't let go of the trauma and guilt I experienced because of Marco's sexual abuse at three years old. I couldn't fix it, but I could give him help if he wanted to fix it. This time he did. Now that he is doing well, I can let go. My story seems to have a happy ending, but I may be in for more pain and disappointment, for I know that many addicts relapse. Marco has had training on how to handle relapses, so if this happens, I will try to trust him to work it through. Now, I hope he can figure out his own life and how he wants to shape it. As he becomes less in need of a parent, I hope we can develop an adult relationship and enjoy each other, not just for moments here and there, but for the rest of my life.

CHAPTER 9

Extending Family

From the time in my late thirties when I saw I might remain single, I was interested in alternatives to the nuclear family. I tried communal living, maintained a network of close friends, organized holiday dinners for friends without family nearby, and more. Now, forty years later, examining my own and Marco's experience with adoption has led me to propose a new kind of extended family integrating members of adoptive and biological families. Such a vision is different than the current ideas about, and practice of, open adoption. Open adoption can include any sort of even minimal contact between adoptive and birth parents and does not necessarily involve meeting in person, and if so, it may be only a few meetings when the child is a baby. Too often, adoptive parents cease to want ongoing meetings, but sometimes it is a birth parent who cuts off contact. Too many adoptive parents still have the model of a nuclear family in their heads and feel threatened by an alternative. In an extended family formed through adoption, however, meetings with a birth family may be only occasional, similar to those in many biological extended families. This conception need not threaten the adoptive parental responsibility. Birth parents may find it painful at first to have contact with the child to whom they gave birth but are not raising. Over time, however, they can come to cherish continued contact with their child. Research demonstrates that ongoing relationships between birth and adoptive families improves the experience for both, and for the adoptee.[1]

As in any extended family, some children will be closer to individual grandparents, aunts, uncles, cousins, and siblings than to others. And some extended families will be closer or larger. An extended family linking biological and adoptive kin, in most cases, would not involve shared parenting. Adoptive parents would be the parents; birth parents would be more like aunts or uncles unless all parties wanted to change this.

My first visit to Louisiana, the one with Marco described in chapter 1, had been cut short because of my brother Mickey's impending death in North Carolina. Knowing that he had less than a week to live, I wanted to participate in deathbed home care with our sister, and with Mickey's wife and his grown son and daughter. Mickey had been a caretaker, and I wanted to help care for him. On the plane to Charlotte, I reminisced about my visit with Marco's biological families and how they compared with mine.

I realized that the values and lifestyle differences between my younger brother and me were as great as those in Marco's birth families. My brother excelled at sports and popularity; I excelled in school. He played football in high school and college, whereas I liked individual sports like swimming and hiking. My brother got an MBA; I a Ph.D. in sociology. In the 1960s, his politics moved to the right, mine to the left of those of our moderate parents. He went with the army to Vietnam; I protested the war. He remained a devoted Catholic; I left the church. In his late twenties, he married and became a family man, devoted to his wife and two children; I remained single. He tolerated his job as a salesman for a major chemical company; I loved my career as a professor.

Because we were so different and because my brother was socially and politically conservative, I appreciated his approval of my nontraditional family. He had given me much by being the primary caretaker of our parents in their declining years and by being so accepting of Marco. I was glad that I had been able, at times, to support him. When he had a major conflict with his daughter in her twenties, I helped them resolve their issues. Despite our geographical dispersion and different life choices, my brother and I visited regularly. We were family.

◆ ◆ ◆ ◆ ◆

To attend the funeral of his beloved uncle, Marco rented a car and drove seven hundred miles from New Orleans to Charlotte, passing through

Mississippi, Alabama, Georgia, and South Carolina. In Alabama he was stopped by a state trooper. Marco knew he wasn't speeding and presumed that this was yet another instance of being stopped for "driving while black." The trooper wanted to wait for a backup so that they could go through Marco's belongings. Marco was furious, but he knew he couldn't show his feelings.

"That's fine with me," Marco said in his most respectful manner, "even though in California you would not have the legal right to search my car without probable cause. I'll wait even though now I'll probably be late for my uncle's funeral."

After another ten minutes, the trooper released him without a search or a ticket. When he was alone, Marco called me on his cell phone and started to cry as he related the incident. Tears began to cloud my eyes, but I was also angry that Marco had been humiliated, something no one in our white family had to endure. Marco told no other family member about his experience, but I'm sure it didn't help the discomfort he already had about his place in his adoptive family.

I loved having Marco beside me at the funeral, burial, and reception. Our dyad was no match for the families of my brother and sister, but it solidified my claim to family. I could tell, however, that Marco was ill at ease. I was glad that several of my brother's colleagues from work were African American. At least Marco was not the only black person in the room as he so often had been at family events in the past. I knew Marco felt insecure with his cousins. In childhood he had established strong bonds with them, but these had loosened as they pursued conventional lifestyles. They all graduated from college, found good jobs, married, and had children. Marco wanted these things too, but addiction had thwarted his ability to achieve them.

Marco talked about his Louisiana relatives and showed pictures of them, but he seemed to belong to neither family. In the Louisiana family he fit in by ethnicity and temperament, and they too struggled with addiction, but he had not grown up with them. I wanted him to be an integral part of both families. I wondered whether if I had opened the adoption when he was younger, it might have been easier to work for such integration. I wondered what I could do now to help him bridge the gap. I decided to return to Louisiana by myself to further explore my impressions and feelings about these families in a more relaxed manner. I also wanted to learn more about the cultures of his Louisiana families, which might help me better relate to them. Could they become family for me too?

As a first step in my cultural education, I decided to attend a five-day workshop on Creole and Cajun culture at the University of Louisiana at Lafayette. It was a safe and easy step. The workshop was informative and fun and made me feel more at ease in Louisiana. I enjoyed the local food and learned that zydeco music evolved from Creole blues. I'd always liked indigenous music, but zydeco meant even more after I learned that the accordion, a central instrument in zydeco, was introduced by German immigrants, making me with my German background feel more connected. Zydeco was played with an accordion, a fiddle, and percussion, often along with a rubboard (fashioned like a washboard), guitar, and sometimes a saxophone. Especially memorable was a Saturday morning breakfast and jam session at Café des Amis in Breaux Bridge, a small town near Lafayette.

On my own, I went on a swamp tour, starting on Lake Fausse, part of the great Atchafalaya Swamp. As the boat drifted along, I observed the lovely, tall cypress trees with delicate Spanish moss draping over their branches. The huge roots of the cypress were half out of the water and on this sunny day created beautiful reflections, providing a much more upbeat impression of the swamp terrain than my original view of the Atchafalaya from high on a bridge on a foggy, rainy day. This sunny, cold day in early December was perfect for bringing out the alligators to bask in the sun. In addition we saw red-ear turtles, a bald eagle, great blue herons, and a white egret. I understood why people loved this landscape.

As part of the workshop, we visited the university's archives of Louisiana author Ernest Gaines, whose novel *A Lesson before Dying* I admired. I learned, too, that Louisiana Creoles and Cajuns are so committed to their culture that, before Hurricane Katrina, they had the lowest emigration rate of any state in the Union.[2] This insight helped me understand why Marco's families expressed no interest in visiting California.

The biggest gratification was when the excellent coordinator of the workshop, a lively middle-aged white woman, revealed in the first-day introductions that she had grown up in Hastings, Reggie and Louise's hometown. She didn't know either of them, but her older brother knew Don, Marco's white uncle. When I told her the story about Marco's delivery to the San Francisco airport, she said, "That must have been Dr. Brousand. He was my ob-gyn too, and he's still arrogant, racist, and sexist." I replied, "Yes," remembering how he treated the black waitress when we had met in the San Francisco airport.

All the years of fantasizing about Marco's origins melted away. I was here in Louisiana talking to someone who knew the doctor who had delivered Marco, the doctor whom I had met ever so briefly in the airport: a Louisiana woman who confirmed my experience of his racism. Ironically, if Dr. Brousand had been less racist, he might never have sent the baby so far away, and I would not have raised Marco. I loved those years being a parent when Marco was young, but if he had stayed in Louisiana maybe he would not have been as lost as an adult. I quickly realized, however, that in all probability, he would have become an addict no matter where he grew up.

Before and after the workshop, I went to visit the relatives I had met previously. These second meetings sometimes altered my first impressions and at other times reinforced them, but all deepened my personal and cultural understanding.

Upon arriving in New Orleans, I had rented a room in a B&B for a few days. Jeanette, the current wife of Marco's father, Reggie, picked me up there and drove me across a bridge, out to her townhouse, the same one I had visited two years before, which she still shared with Reggie and her teenage son.

The urban setting was drab, with no trees, no sidewalks, and many unpaved streets. Their row of new, attached townhouses bordered a huge concrete shopping plaza, and we heard the noise of cars whizzing by on the nearby freeway. A high-rise project loomed around the corner, with unemployed men hanging out in front. Reggie had said he preferred Hastings, his hometown, with its greenery and small-town flavor. I could see why.

During my short visit, Reggie remained as quiet and elusive as he had been two years before, but he showed off his cooking skills by serving me excellent leftovers from Thanksgiving. I especially liked the gumbo, stuffed green peppers, and sweet potato pie. I spent more time with Jeanette, a friendly, outgoing woman. In the past two years, she had become my main contact with the Creole Louisiana family through our ongoing e-mail exchange.

As we ate and talked, a large TV, turned on low, dominated the room. Jeanette often glanced at the movie, commenting on the plot of a film she had obviously seen before. I found this distracting, something I would never do with guests, but the neat, cheerful room, decorated with lots of photos and Reggie's drawings, impressed me with its family mementoes and sense of order. I got used to the television and found myself ignoring it just as Jeanette did. She also showed me poems Reggie had written to keep him sane while in prison. It struck me that Marco also liked to draw and had once won a prize for a poem he wrote.

Jeanette took me back to the B&B and picked me up the next morning to take me with her to Sunday mass at her African American Roman Catholic Church, St. Augustine, in the famous Tremé neighborhood in New Orleans. Going to church would give me further insight into Creole culture and permit me to spend more time with her. Jeanette usually sings in the choir, but this morning, she sat with me, for which I was grateful. The large congregation of three hundred or so was overwhelmingly African American with many shades of brown, and only five or six people who looked white but might have been Creole too. Sitting with Jeanette, I was aware that her dark skin must contrast strikingly with my own pale skin. Like Marco when he was in a predominantly white public space, I wondered if everyone was looking at us. But Jeanette made me feel comfortable at this service, which was so different from the masses I had attended at predominantly white Catholic churches.

The congregation of families and single adults dressed casually, not wearing the fancy attire and big hats with which I had stereotyped African American churchgoers. The mass was in English, with lots of participation and singing by the congregation. A band played gospel music and accompanied the choir. I loved it—so much more engaging than the staid, quiet Latin masses of my childhood.

As he began collecting gifts and money to give to prisoners during the upcoming Christmas season, the black assistant pastor—not the white head pastor—gave the sermon. He preached about helping those in need and noted that once in a larger congregation of three thousand African Americans he had asked how many had experience with prison themselves or through family or friends. Ninety-five percent raised their hands. Wow, I thought, this church also has progressive politics, not like the Catholic church my brother and his family attend in North Carolina where the priest preached against Obama's acceptance of the right to choose abortion.

After church Jeanette and I drove through the historical black neighborhood of Faubourg Tremé. Along with the lovely streets of renovated townhouses, we saw neighborhoods still in ruins more than six years after Hurricane Katrina, along with a few high-rise projects and areas where wreckage had been bulldozed, but nothing yet rebuilt. I learned later that at its founding in 1841, Tremé was the first urban African American residential community in the country. Residing there were free blacks and mixed-race Creoles and Native Americans. Some say that jazz was born here and only moved later to the nearby French Quarter. In films and documentaries, I had seen wonderful Mardi Gras parades in Tremé and other, smaller parades with musical accompaniment for weddings and funerals.[3] But on this winter day, the streets were quiet.

In one of the desolate areas, Jeanette chose a cozy diner for lunch with delicious fried chicken and waffles. Over our meal, we talked more personally. Jeanette revealed that she didn't know if she and Reggie would make it, as she was fed up with his drug use. He would stop for a while, but then resume. She thought he traded food stamps for drugs. Moreover, she learned recently that he had a two-year-old son—his fifth child—in foster care in a city sixty miles away. He didn't believe that the child was his, but DNA testing had proved otherwise. Jeanette would have been willing to overlook his infidelity and help raise the child, but it now seemed unlikely that social services would release the child to him because of his prison record, unemployment, and history of substance abuse. Hearing her story, I was thankful again that Marco had not yet had any children.

A year later, Jeanette called me in California to say she had had a small stroke and spent two nights in the hospital. Reggie never came home or called and never visited. That was the last straw for her, and she kicked him out. But a year after that she called to say that he had changed and they were back together. I wondered why she thought his change would last. Her roller-coaster with Reggie was so like mine with Marco. I was glad to have this confidante in the Creole family, and we kept up an intermittent correspondence for years to come.

After the workshop, I rented a car and drove to Hastings. Reggie and Louise, Marco's birth parents, were born and raised here on different sides of the tracks (literally), meeting at age twenty-five and age seventeen respectively. Louise came from a stable but poor white family, Reggie from a more complex black one.

I remembered my previous visit to this small city surrounded by farmland with rice as the primary crop. The main street with its chain stores and restaurants seemed like any town USA. But as I crossed the tracks, I encountered another world, a predominantly black neighborhood with only a few small stores. I entered Hastings with trepidation because of stories Marco told me from the four months he lived here with Reggie and because of a story about Hastings I'd read in a national newspaper. The story reported on eight unsolved murders of young women, both black and white, from poor parts of town. The paper described Hastings as a "stopping off point for drug traffickers along Interstate 10, a town with a 'thriving crack trade.'" I was relieved that my first impression was more positive as I surveyed the houses from the car window and saw few of the derelict ones described by the newspaper.[4]

To be sure, the one-story homes looked small. A few were boarded up or had junk in the yard, but most were painted, sporting nice porches and well-tended yards. On this sunny, cold afternoon the streets seemed peaceful and empty. But I remembered what happened when I came here with Marco and Reggie two years previously. On our way to visit Reggie's half-sister, Dottie, we had stopped at the home of Reggie's friend so that Marco could say hello. They asked me to wait in the car. Bored, I grabbed my camera and got out to document my visit. All at once, Marco ran out of the house, took my arm, and ordered me back in the car. "Do you want to get yourself shot?" he asked. "A white woman with a camera is seen as a threat, someone from law enforcement, or a federal agent." Maybe the newspaper article had it right.

Today, I drove in again to see Dottie, the one with the small house and the beautiful oak table. As it was Sunday, I had asked Dottie, too, if I could accompany her to church. Hers was an evangelical Protestant service in a converted community center outside of town with only thirty-five people attending. Again, I noticed the racial composition of the congregation—primarily African American with a sprinkling of whites. As in the Catholic church, an excellent band of piano, drums, and banjo accompanied a few accomplished singers and the congregation. But this service was quite different from the Catholic one and made me uncomfortable. The minister, a short, heavy-set, black man with a pugnacious style, gave a long, rambling sermon that thundered against the sin of extramarital sex. He seemed to blame women for seducing men, only seldom mentioning that men come on to women. At one point, he asked someone to open the door. He then screamed: "Let natural sense and reason be gone. Only spirituality lives in here."

I asked myself: Didn't one need reason to resist the temptations he was preaching about? My feminist sensibilities bristled. I didn't trust him, and I wouldn't have left him around any young women. Unlike the Catholic church service, which drew me closer to Reggie's wife, this evangelical one distanced me from Reggie's sister.

At one point, I noticed that Dottie was crying. After the service, she told me that she cries every Sunday, so grateful is she that this church has saved her from the curse of addiction. All I could think about was the strong tendency to addiction in this family. She told me that only once had she succeeded in getting Marco to accompany her to this service. It certainly didn't appeal to me, and I could see why it hadn't appealed to him. My rationalist upbringing had made an impact on him and for that I was glad. But if any church could have helped cure Marco's addictions, I would have gladly accepted his conversion.

After church Dottie took me for a brief visit with her father, Marco's grandfather. Half-siblings Reggie and Dottie were only two of his twelve children by a variety of women. Since my last visit, their father had moved back to Hastings and married a fifty-year-old woman, who Dottie said was into the street life and exploiting her father, now in his late seventies. We found him hanging out on a corner in the black part of town where he spent the day chatting with old friends who wandered by. What a dresser that man was! Perhaps Marco had inherited his sense of style, even though he had known nothing about this grandfather until a couple of years ago. His outfit consisted of light rust overalls with the same color leather jacket, an orange shirt, and a large, intricately carved silver pendent on a silver chain. I wanted a photo, but when I asked him to smile, he refused. "I don't want my missing front teeth to be recorded," he said.

As I dropped off Dottie at her house and drove out of town, I thought it would be hard for me to relate to Reggie's family on a long-term basis. There was too much pathology. My reaction was not just because they were African American. Later, back again in Lafayette, I would meet Jeanette's mother and would have a different feeling about that black family.

While driving to Lake Charles, I wondered how I'd feel this time about Marco's white birth family. I realized that my way of visiting a number of different relatives who live within driving distance of each other was how I often visited my kin in the East.

Lake Charles is a city in southwest Louisiana, near the Texas border, which sits beside a beautiful lake. The city hosts several casinos and huge petroleum refineries that dominate the horizon. Within view of the lake, hefty towers and massive, aboveground pipelines hover over the landscape, turning what could have been a scene of natural beauty into a futuristic nightmare. At night, the refinery fires light up the sky, spewing out deadly pollutants. Later, I learned that the towers are used to heat up the crude oil and to distill a number of products, including gasoline, diesel, kerosene, and alcohol. The pipes separate the products and move them around, some going more than thirty miles to tankers in the Gulf.

I was here to visit Louise, Marco's birth mother. I had learned from Marco that she had lost her job at Target. I never discovered whether this was due to the economic downturn or if the employer found some fault with her, real or concocted. With none of the discomfort she displayed toward me during our first visit two years earlier, Louise was friendly, chatty, and open, revealing more about herself. Perhaps she assumed or knew that Marco had told me about their drug use together. That I was here for a visit meant I had not rejected her. I was interested in her because I was always thinking about the possibility of extending family.

She told me that after her firing, a friend hired her in the family's sausage shop. Six months later, however, the family decided to close down in the off-season and no longer needed Louise's help. For the past year and a half, she had worked in the furniture store owned by her sister and brother-in-law. Louise confided that her husband was in jail for a year. She didn't know whether she would get back together with him after he was released. She told me that he was an alcoholic with a long record of alcohol-related offenses.

Louise revealed little about her own history of addiction and I didn't want to say how much I knew, fearing that it might be a betrayal of Marco. She did say that living by herself made her more prone to depression and to using drugs and alcohol. She saw and talked to her children often, but had few friends. Working, taking care of her kids and getting in and out of relationships had left her little time to develop friendships. This was very different from my life, where friendships are central.

Louise is proud of her children and their accomplishments, especially since they were troubled growing up. She had been strict and accepted help from her family. Her son will probably leave the military soon and get a good civilian job. Her daughter will graduate from college in a few months and is engaged to a young businessman in town from a well-to-do white family. I couldn't help reflect to myself that the children she raised had turned out better than the one I had nurtured. Such thoughts could easily lead me to question liberal child rearing and adoption itself, but I resisted going there. I focused instead on how I might fit into these families if Marco wanted to maintain his ties to them.

I continued these thoughts as I drove back to Lafayette to see Don, Marco's uncle, the educated social worker and hunter whom I had so liked during my first visit; he made a different impression this time. I'd thought we had much in common, but on this visit I noticed our differences. When we met for dinner, he brought along the priest with whom he collaborated at Catholic Charities. From their talk, I now saw Don's devotion to Catholicism. I admired his social conscience and work with the poor, but I couldn't identify with his religious devotion. I noticed how much of my time in Louisiana had been spent experiencing and talking about religion. This was also true when I visited my relatives, but was not a part of my life in Berkeley. In my experience, it was easier for believers and nonbelievers to peacefully coexist in an extended family than it was for relatives with different political views. This was why I never initiated political discussions in Louisiana.

I soon realized that Don was a regular guy to whom I could easily relate. He didn't know the novels of Louisiana author Earnest Gaines, but he was devoted to the LSU football team. In the restaurant, he kept peeping at the game on TV and apologizing for doing so. Only when LSU got far ahead did Don relax. He said he didn't watch their games at home because it made him too anxious.

Don laughed when I said that I had been hoping that he would cook me squirrel for dinner. Now he shot deer and liked to eat venison. This season, he had shot three deer during the thirty-seven days he had been hunting. He didn't stalk his prey, but waited in a blind where he sat watching birds and meditating. He said, "Hunting for me is a spiritual experience; nature gets me in touch with God." I could identify with this view of the deity.

I was surprised when Don said that after retirement in a few years, he planned to move back to Hastings, where he had grown up, and would buy a trailer to live in. He will be close enough to visit his three sisters, to continue doing volunteer social work, and to hunt. Don had no interest in travel and seemed to epitomize the Louisiana love of place that the workshop had highlighted at the beginning of my trip.

Although Don and I didn't have much in common, our differences were actually less than those between me and my brother, with whom I had grown up and who shared half my genes. My brother liked sports as much as Don, wasn't an intellectual, and had conservative politics that were more alien to me than were Don's, which seemed liberal. Yet my brother and I were close because of our years growing up in the same family. Don and I met when we were both in our sixties, making it more difficult to forge a bond.

Whatever the potential of our relationship, I saw now what Don shared with Marco. Don, like Marco, was open, sociable, loved the outdoors, and was anxious. But it is hard to separate nature and nurture, since all those traits characterize me too.

Both Louise and Don told me of a recent family Thanksgiving where the extended family all gathered at their sister's big house. Even though they said they were not that intimate with each other, they have a family loyalty and help each other out. I could imagine going with Marco to a holiday dinner with them, seeing them as extended family, one that, unlike mine, contained his mixed-race siblings. I also could imagine joining Jeanette, Marco's Creole stepmother, who celebrates with her son, her widowed mother, and her sister with her husband and their three children.

Jeanette's mother, Gloria, also lives in Lafayette, where she moved to escape from Katrina. She got a good job there and stayed. We went out for a lunch of oyster po-boys and talked. Like her daughter, she was open and friendly, and seemed more grounded. She said about her daughter's husband, Marco's father: "I like Reggie, but he is crazy." I wish I had asked her to explain, but this question seemed a bit too intimate to ask someone I had just met.

At a Thanksgiving dinner with Gloria's family, I might be the only white person there, but both Jeanette and Gloria are so warm and welcoming that I knew I would be accepted. Reggie's family, however, was too broken and dispersed, with too many unresolved conflicts, to assemble at the holidays.

I wish I'd had this vision of extending family when I adopted. Now I was too old and too dependent on Marco's decisions about what kind of ties he wanted with these relatives to initiate extended family ties in Louisiana. In 1981 when I adopted, I wanted to create a communal household as an alternative to the *nuclear* family. Alternative, not *extended*, families were prized by the counterculture in Berkeley, as many of us rejected the values of our biological families and believed we could nurture superior ties based on friendship and community.

Bonds of friendship and community still anchor my life, but as I matured, I saw that they did not replace family ties. Adopting Marco helped me strengthen bonds to my biological family. I now see that I could have forged additional family connections with his biological family. If I had done so when he was younger, both he and I would have had a biological and an adopted family, different for each of us but bringing a balance into our family life. My biological family was his adoptive one; his biological family could have been my adoptive one. This would have benefited Marco. In a 2016 interview with author Andrew Solomon, an interview where I was not present, Marco said, "I wish I could have lived both lives." He was referring to his adopted life with me and to the reunion he had with both birth families. He also said, "When I met my biological families, and spent time with them, my feelings of insecurity were gone."[5] If I had tried to maintain some contact with his birth families when he was young, he might have had this support earlier.

Family is idealized in our culture as it is in most societies, yet we know that often practice falls far short of the ideal. We know that biological families—nuclear and extended—can often involve conflict, anger, splits, disappointments, and tragedies. We know that even within a close extended family, some relationships are more intimate than others. Biological families also have to deal with the impact of mental illness and addiction. Some biological families incorporate class differences, as did Marco's family of birth. Extended families integrating adoptive and birth families will be every bit as complex.

Both Marco and I are closer to some of his Louisiana relatives than to others. Both of us have closer ties to his father's Creole family, me to Reggie's wife (no biological relation) and Marco to his younger half-sister, his father's daughter. I would have thought I would be closer to the white family, which is more like mine in social class and religion, but guilt undermines our closeness—the birth mother's and my guilt that in distinct ways we enabled Marco's substance abuse. Don may feel guilty that he acquiesced in the adoption.

In December 2017 a tragedy occurred in Marco's Creole birth family, but it demonstrated that both he and I do have ties to this extended family. For five years or more, Marco had known that he had a half-sister, Catlin, Reggie's first child, about two years older than Marco, living about seventy-five miles east of Berkeley. Although they communicated on Facebook, Catlin and Marco had never met. Each said they wanted to, but it never happened, largely due to their mutual substance abuse. Marco didn't have a car, but he talked about having a friend drive him. I didn't say anything, but I never volunteered to take him, because Marco told me that Catlin also had addiction issues. I was fine with her life as a lesbian, living with a partner, but I was put off when I heard that she had lost custody of her younger son to a relative, while her older son (who was seventeen or eighteen) lived with her own mother.

On December 15, Marco received a text from his younger half-sister in Louisiana, while I received an e-mail from Jeanette, Reggie's estranged wife. They told us that Catlin had died of a heroin overdose. Marco sank into depression at this news and had deep regrets that they had never met. He showed me pictures of Catlin, and they did look like sister and brother. I again thought that such a tragedy, like the death of his friend Matt, would lead Marco to sobriety. Instead, he drowned his sorrow in more drugs and alcohol. He asked me if I would go with him to a service for Catlin. With no hesitation, I said, "Of course."

Jeanette gave me a phone number for Catlin's mother, Pauline, who told us there would be a memorial service in early January. It would take several weeks for the police to release Catlin's body because they needed to do tests to verify the cause of death. Pauline seemed pleased that Marco and I wanted to attend. She wanted to set the date and time for the service so that Reggie could come. Even though Pauline hadn't seen Reggie since Catlin was four, she wanted him there. This seemed possible, since Reggie now lived in southern California with a new girlfriend. He said he would come if they could schedule around his work hours. Finally, we heard the memorial would be on January 9 in the early evening, the day before Marco's birthday.

Marco was excited about seeing Reggie again and invited his best friend, Paul, to come to meet his father. On the morning of January 9, however, we heard that Reggie couldn't make it. Marco was upset and angry. "That man is such a fuckup," he swore, and was more determined than ever to attend as the only representative of his father's family.

I was happy that Paul still agreed to come, since he was a good driver and I didn't want to drive on the freeway in rush hour in the rain and then back in the dark. Paul brought his five-year-old son, since his wife had to work. Isaiah and I sat in the back seat talking and playing, which was a nice distraction.

The GPS found the church building in a desolate part of town near the airport. The small structure looked like a house made of unpainted boards. We arrived at 5:30 p.m. to find the building dark and locked. Did we have the right place? The address was correct and the name of the church was on the front, but there were no signs of life. After just a few minutes, fortunately, cars began to pull up. The minister arrived, opened the church, put on the lights, and retreated to a back office to write his eulogy. Pauline got out of another car and embraced Marco. Their first words were about their anger at Reggie. Marco helped Pauline and several of her friends carry pans of cooked food and groceries into the spacious kitchen and dining area. The church itself was simple but attractive, with pews and an altar made of polished wood, not like the much more impressive Catholic church in New Orleans, but much nicer than the community hall outside Hastings, where I had attended the evangelical service.

Pauline said that she needed to go pick up her husband, who belonged to this congregation, and to finish a few more preparations. Another woman and I volunteered to organize the food. A friend of Catlin's brought decorations for the church—an enlarged portrait of Catlin, purple and white flowers, and purple paper streamers. We didn't need to be told Catlin's favorite color.

Other Creole relatives and friends slowly filtered in, seeming to know that the service would start late. Pauline told us it probably would not start until 7:30 as they would wait for Catlin's son, who was driving back from his job in Fremont in heavy traffic.

Marco discovered that most of those coming into the church had roots in Louisiana even though they had lived for some time in California. Many were from Hastings, Pauline's hometown as well as that of Marco's parents, Reggie and Louise. The men wanted to talk with Marco, who had been in Hastings more recently than they had. Some knew Reggie and wanted to see where Marco fit in. These discussions made Marco feel at home.

In the kitchen, I found that the other white woman called herself

Catlin's mother-in-law. Her daughter had been Catlin's partner until they separated a month previously. She showed me photos of the couple. Her daughter, she said, was so distraught over Catlin's overdose that she could not bear to come to the service. While the Creoles gathered in the chapel, we white women worked in the kitchen. It felt right, as compensation for all the years black women had worked in white women's kitchens, even if never in ours.

We put the prepared hot dogs in a spicy sauce in the oven to keep warm and stashed the potato salad in the refrigerator, along with the soft drinks and lemonade. I was relieved that there was no liquor, wine, or beer. We peeled and cut fruit and piled it on platters. We cut up the lettuce and vegetables from Safeway for salad and placed the bowls next to the bottled dressings. Marco and Paul helped us open the huge cans of baked beans and pour them into a large pot to heat. There was enough food for one hundred and fifty people, but there were only about fifty present. Later, Marco commented to Paul and me that the food in Louisiana would have been better, even at a working-class church. I agreed.

Finally, the service began, two hours after we had arrived. Pauline asked Marco to sit in the front family pew with her and her husband and grandson. I was asked to sit in the next row with close family members, a request that pleased me. Other relatives and friends filled in behind. The African American pastor gave a short eulogy. I don't think he knew Catlin, but his words and sentiments moved me; he was nothing like the ideological evangelical in Louisiana who had offended me with his antiwoman cant. The pastor then announced that five people from the audience could come up to speak, but only three did so: Pauline's husband, who was Catlin's stepfather; the mother-in-law who I'd met in the kitchen; and a young woman friend of Catlin's. They all spoke of Catlin's beauty and generous spirit that had been damaged by addiction. I felt sad about her wasted life. After the service, Marco tried to talk with Catlin's son, but the young man looked sad and did not say much. On the way home, Marco vowed he would keep in touch with his nephew.

I was glad that I could participate for a short while in a world very different from mine—glad that I could be part of Marco's Creole family. Some months later, Marco told me that he texts or talks almost every day with his younger sister, Caroline, now twenty-nine, married with five children, and living in a small town in Louisiana. Together, they are trying to locate another half-brother, who they've been told lives in Texas, and a much younger brother who they think was adopted in Baton Rouge. Marco told me that Caroline doesn't have the addiction gene and that she and Jeanette keep the family connected.

Recently, Jeanette, now divorced from Reggie but in friendly contact with him, sent me an e-mail inviting Marco and me to her wedding to a new man. Her words moved me. "You and Marco will always be a part of my life," she said, "and Marco will always be my son." In summer 2020, Marco and I will attend a Trimberger family reunion on the East Coast. Perhaps Marco will invite someone from Louisiana to join us. If he has a wedding someday, I'm sure he will invite Louisiana relatives.

These family relationships illustrate the importance of extended family ties in adoption, beyond just contact between the adoptee and his or her birth parents. Reggie has disappointed Marco, but his Creole sister and stepmother have gathered him into their family circle. This past Christmas, his first in sobriety, Marco took the initiative to text holiday greetings to some of his white/Cajun relatives too. Perhaps they will re-enter his and my life in the future.

Extending family ties, integrating biological, adoptive, and chosen kin, brings nature and nurture together. Adoptees can have a wider network of kin with whom to identify and relate. Their identities can be more secure. Birth families don't lose their kin to adoption. Adoptive parents gain access to cultures and people they would never have experienced, enriching their lives.[6] In our case, we have concocted a delicious gumbo.

Afterword
The Creole Son Speaks
Marc Trimberger

My mom asked me to read this book to be sure I was comfortable with what she wrote and that it was accurate, but I couldn't stop turning the pages for I was reliving much of my life. Seeing the complexity of my life on the page was good for my sense of self. But it also brought up situations and feelings which I had blocked from my memory. I think this is a common occurrence for those of us who have had painful or stressful experiences. At times I felt ashamed or embarrassed by some of my life events, but I reminded myself that every part of my life has made me who I am. I am the result of my positive and negative experiences, although I don't like to use the word *negative* because I believe a person can find something good if only a lesson learned and not to be repeated.

For me, being adopted and growing up mixed-race with a white mother in a single-parent household never seemed strange or unusual. I can't remember feeling like I was different or having many issues with this situation, even though I was aware of it from as far back as I can remember. It just didn't create many problems for me, especially because there were several black boys with white single moms in our neighborhood. If anything, looking back on my childhood, adoption was a blessing, given what I now know about the racial tensions present in my birthplace in Louisiana. I felt I was lucky to have two birthdays, the second one being my arrival day. Being adopted gave me an edge, a little more character, so I was not just like everyone else. Only once in my first eleven years, up until junior high, did I experience a problem with being adopted. I was involved in an altercation with a kid who made fun of me for not having a real mom. It didn't make sense to me, because my adopted mother was then the only mom I knew.

I never thought about the so-called problems created by adoption until I was an adult and watched the film *Losing Isaiah*. I never felt lost or out of place. Except for that one time, I never was teased or picked on or excluded because I was adopted. Coming to terms with being black was harder. Starting in junior high, I realized that I didn't understand ebonics, the black talk used by some of my peers, because I hadn't heard it earlier. But I found it easy to pick up. A few times, I would forget that my mom and other white friends couldn't understand it. Mostly, I could switch languages easily between my white and black worlds.

I felt a lot of tension, however, between my desire to be one of the cool "niggas" in my hood and being more comfortable with a mostly white group of friends who, like me, had attended private elementary school. Even though some of my new black friends often tried to test my gangsta by picking on me or trying to bully me, I felt the need to be accepted by them. I realized my teen life was more difficult than that of my white friends. These tensions stayed with me until my thirties. Even when I lived in middle-class Berkeley or Oakland, I would still dibble and dabble in the hood.

I was very eager to meet my biological family. Being able to meet my dad was more important than anything, because I always wanted to have a dad. I remember asking my mom on numerous occasions if she had a boyfriend and if she was going to get one, mainly because I hoped if she did find a man then I could have a dad. If I hadn't had a fairly large circle of friends, some of whom felt like brothers or sisters, I probably would have wanted to find my biological family in order to find siblings. As it was, I mainly wanted a dad, but it was wonderful to learn that I had six half-brothers and -sisters, all mixed-race or black.

I had the desire to meet my biological mother, but I was a little apprehensive because I love my mom, whose book you're reading, and I felt I hadn't shown her how absolutely incredible she is, how much she means to me. Even though my mom had arranged for the reunion, I didn't want her to think for a moment that someone could take her place. I was scared it would hurt her if I connected with my biological mother. Because I had caused my mom so much stress as a teenager, and because I knew I was in no way living up to my potential or pursuing a lifestyle which reflected the way I was brought up, I feared she might feel I didn't care as much as I knew I did. This was my greatest fear as an adopted child. I was fortunate to be able to discuss my fear with my friends, who reassured me that my adopted mom was my mom, period.

I remember meeting my birth mom first, then my brother, and then my father, who was the person I wanted to meet most. I appreciate that my birth mom had told her other children about my existence, that she was so welcoming, and that she helped me find my birth dad. That first day when he came to my birth mom's house, we played dominoes, talked, and drank Tanqueray.

Louisiana was a welcome change to the fast-paced city life of the San Francisco Bay area where I was raised. I remember when I first arrived at my biological mother's mobile home in the middle of a sugarcane field. There were no sidewalks, and the nearest store was a forty-five-minute walk. I was used to living in a city with busy streets, houses all bunched together, and a store on almost every corner. But I liked being in the country. The food was also different and interesting.

I remember the first time I looked in the fridge. Instead of finding ground beef, pork, or chicken, I saw deer sausage, squirrel, and duck. I couldn't wait to get a taste of these meats that I didn't even know people still ate. I thought all those stories of eating frog legs, raccoon, and squirrel were just stories, or things people said to make jokes about the South. When I was living with my birth mom, I ate the sausage, squirrel, frog legs, and duck. Later, my dad cooked coon, rabbit, turtle, and armadillo. I liked them all, but my favorite was squirrel.

I loved the energy people put out in Louisiana, with friendly smiles all around. I was always worried about racism in the South, but I didn't get that type of vibe from people. I'm sure there are still racist thoughts and practices in Louisiana, but I experienced more in liberal California.

Meeting my birth family was very important. It was the key to my being comfortable in my own skin. It gave me a new sense of self and made me feel more O.K. with my life. I had been a little tough on myself about where I was headed and how little I had achieved, since I obviously possessed the skills and ability to do so much more. I definitely looked down on the man I had become. I was so different from my cousins, the almost perfect children of my mom's brother and sister. It was like they were programmed to be these perfect kids. I was uncomfortable that even my younger cousins were more educated and mature than me. I wished I could show these relatives that I was able and I didn't want to be a negative reflection on my mom. I knew that deep down, even though she never spoke it, my mom felt responsible and she felt embarrassed or shamed by the way I had turned out. At least so far. But I felt stuck in my lifestyle.

When I reunited with my birth family, I found out how similar I was to my birth parents. This lifted a big weight off my shoulders. I could now see that I was O.K. I wasn't just a fuckup and I had a chance to become a son who, with some effort, would make my mom proud. Although it's been a slow process, I am no longer stuck.

After I first met my father, who had been recently released from jail, we took a trip to meet my younger sister, my father's fourth born. We have very similar features, and despite an eight-year age difference, we call each other *twins*. Caroline always loved our father, but for her whole life, he was in and out of jail and she resented him for it. She didn't really have a connection to her father's family, which she wanted. I was looking for that same connection. I feel this is the reason we became so close. When I returned to California, we started communicating on a daily basis. Now, more than eight years later, we still text or talk every day.

I don't feel that my being adopted or my childhood or the way I was raised are in any way the source of my long and very destructive journey of addiction, from which to this day I struggle to free myself. Before learning about a genetic predisposition to addiction, I felt that it was likely the outcome of a desire to fit in with and be like the kids and adults I thought were cool. It is possible that being adopted led to some insecurity that increased my need to be perceived as cool.

I first became curious about drugs when I found myself thinking that people who smoked cigs were so cool. This was a complete 360-degree turn from the first six or seven years of my life, when I remember telling people that smoking was nasty and bad. When I tried cigarettes, starting at age eight, I couldn't get enough. I was smoking anything I could to pacify my desire to feel cool like the adults I saw smoking. I used grass, oregano, leaves, anything that I could roll up even in plain paper, as long as I could light it and see smoke. No one I knew did it, and there wasn't any smoking in my household. It was my own obsession. After reading the science in this book, I see that my reaction to nicotine, the first drug I used, was more intense than that of kids who didn't have this genetic predisposition. It wasn't just that I wanted to be cool.

When, in the later years of high school, drugs like weed, cocaine, ecstasy, and meth became part of the arsenal to feed my addiction, it was not about racial identity, because few black kids I knew tried any of them. It was affluent white kids who were drug users and sellers. As early as junior high, if you wanted some weed, you would get it from white kids. My being a black with many white friends meant I always had access to drugs and made black kids look up to me. I remember my black friends coming to me with what they had bought, thinking it was weed, and I would laugh when I realized that these cool kids I wanted to be like didn't even know weed from oregano. They were getting played.

I had indulged in the use of many drugs, but it wasn't until I reached my mid-twenties that I actually understood what people mean when they speak of addiction. Before this I had no problem picking up and putting down drugs. I didn't feel they controlled me. I didn't have to have them and never stressed about needing them. But at twenty-four when I started using crack cocaine, all this changed. I no longer was in control. I became consumed in the endless effort of getting the money I needed to get my issue. Using became my only focus. I couldn't think about anything other than what I needed to do to get my next hit.

I sometimes wonder what meeting my birth family might have been like if I never had used rock cocaine, because it just so happened that my biological dad and mom both had struggled with the same drug throughout their lifetimes. And what happens when addicts get together? In my experience, they use, and that's just what occurred during the time I spent getting to know these very important people who gave me life. We hung out like we were longtime friends on an endless mission to stay high. "How great is this," I remember thinking. "I get to meet my birth family and I'm getting high at the same time." Everything was just perfect. I couldn't believe how cool it seemed to have my cake and eat it too. But the reality was that there was nothing cool about it. Even though over time we all realized that this was not a good situation, we still kept it up.

I returned home to California and swore that the next time I touched down in Louisiana things would be different. But when I returned to the bottom of the map, there we were again. Same people, same drugs, and same destructive cycles. I feel responsible for my biological mother's relapse. My dad, like me, had been actively using most of his life, so I didn't feel responsible for him.

On my third trip to Louisiana, I lived for a while with my father, his new wife, and her son in a suburb near New Orleans on the west bank of the Mississippi. There I was introduced to Mardi Gras, drive-through daiquiris, and pickled pig lips. Open containers of alcohol were legal, and you could buy liquor twenty-four hours a day. This was a big change from the 2 a.m. last call in California. I now understood why New Orleans was such a popular place for people to vacation and party. Now that I'm living a sober life and am a member of the AA fellowship, I can relate when others tell stories about when they went to party in New Orleans and were lucky to live through it without drinking themselves to death. For an alcoholic, New Orleans, Louisiana, can mean suicide.

It has been about eight years since I spoke to or saw my birth parents. Even though they have tried to contact me, I made myself impossible to reach. The guilt of disappearing on them seemed easier to deal with than that of feeling responsible for being a major factor in the family's demise. Back when I was in contact with them, all this riding on my shoulders gave me more reasons to self-medicate, for I had more stress to numb. What better way to not feel than to use more drugs. That's when things went swiftly downhill.

I met a girl who like me used crack, and I went all out. A few years passed, and if I didn't already have enough self-hatred to fuel my drug use, life hit me with another blow. My friend and brother, who was part of the family I created to fill the need for siblings, was murdered. Now I lived in a dark cloud filled with coke smoke and I found myself at what might be called my rock bottom. Hour after hour, day after day, I used. I refused to deal with anything but finding drugs. Life became a slow suicide by way of crack consumption and denials. Living in an apartment or a motel turned into living in homeless camps, and feeling O.K. meant being high. Eventually, I alienated my friends and related only to a small group of users and dealers. I was high or in misery.

Finally, I was able to see the zombie I had become. After six or seven years of neglecting myself and completely submitting to my addiction, I knew I had to change. I owed it not only to myself, but to my biological family, to my friend who died, and most of all to my mother Kay. With a great deal of effort and by distancing myself from people and areas which were home base for my drug use, I was able to eliminate most hard drugs from my life and cut down on my marijuana use. I renewed contact with old friends who never used hard drugs, and who now gave me a lot of support.

I am amazed at how much my mom was able to explain through her research and to learn she didn't hate me for my lifestyle, one completely opposite of what she wanted for me. Reading research showing that some things are inherited eases my pain at letting her down. Even though it made me relive the disturbing abuse in my younger years, reading this book led to an uplifting feeling.

But getting off drugs was not easy, and my alcohol consumption increased. I often relapsed. After my older sister passed away from her addiction to drugs, I hit bottom again and decided to get some help. My mother found me a bed in a rehab. I left Berkeley, headed to the outskirts of Petaluma in Sonoma County, to Olympia House. In my thirty-five-day inpatient program, I was able to abstain from using drugs and alcohol. I learned about relapse prevention, trauma, anger management, cognitive behavioral therapy, dialectical behavior therapy, meditation, and many more tools that were essential for me to get my life back. I worked with therapists and counselors and other addicts to start building a foundation for a new life.

After leaving the inpatient program, I attended a seven-week outpatient program at Olympia House while moving into an SLE (sober living environment) house in Marin County. There I started attending AA meetings regularly. I got a sponsor who has guided me through the Twelve Steps. As a result of this work, I have made many changes in my life. I no longer am homeless. I get to sleep under a roof, and I have a bed. I got my driver's license back and I own a car. Recently, I got hired at an upscale grocery chain. I have also rid myself of the warrants which have haunted me through all my years of addiction. I am so very grateful for the help of my mom, Olympia House, my SLE, and the Marin County AA fellowship for a chance at a new life. I have a way to go, but I now have positive energy when thinking about my future.

APPENDIX

Implications for Adoption Theory, Practice, and Research

My story should in no way be seen as a rejection of adoption as an institution despite the challenges it presents. At its worst, adoption exploits poor people, including stealing their children against their wishes.[1] At its best, adoption provides children a home and nurturance when their biological parents and extended biological families are not able to care for them. Adoption gives adults who want to parent children to raise. But ongoing adoption reform is necessary, even in the best of circumstances.[2]

Like 87 percent of adoptive parents in a national survey, I would make the decision again to adopt,[3] although knowing what I know now, I'd do many things differently. Adopting, no matter what the reason, requires self-examination as to what a parent has to give and what they hope to gain. I hope my story and research will help with that examination. My experience and research have altered my intellectual point of view, from "nurture is everything" to "nature is significant." While nurture may be less important than nature in determining the adult outcomes of adoption, nurture creates significant attachments that usually are as, or more, important to the adoptee than biological similarities to people with whom the adoptee has not grown up.

THE IMPACT OF NATURE

A primary point of my book is that we must give *nature* more emphasis, and specifically bring a knowledge of behavioral genetics research on adoption into adoption culture and practice. Neither those seeking to adopt nor adoption professionals can any longer ignore genetics. Nurture is not everything. Children do not come into the world as blank slates. No longer can an adoptive parent say, "This child was made for me," an attitude that seems to ignore the child's biological roots. No longer can we say that love is enough.

No longer, either, can we blame adoptive parents when their adopted sons and daughters have troubled lives. Nor can biological mothers be blamed for most of a child's problems. My research leads me to reject the idea that nurturing by parents—biological or adoptive—is the most important determinant of adult outcomes. Rather, I see genetic heritage interacting with the environment, both inside and outside the family, as more significant in explaining the life trajectory of all children.

My perspective can explain why many adoptees come to lead satisfying and productive adult lives while others have more difficulties. Of course, the prenatal and preadoptive environment, especially that in foreign orphanages or in abusive families, will have a big impact on adoption outcomes. Here, too, genetic inheritance can play a role in how an adopted child later handles early adversity.[4]

Most counterintuitive for me was the behavioral genetics finding that differences between parents and their adopted children increase over time. By the time they are young adults, adoptees will be more like their biological than their adoptive parents, even when there has been no contact with the birth families. This finding helped me understand why my relationship with Marco changed so much over time and why it has been so hard for me to accept his adult life. It points to the need for more support and education for adoptive parents of teenagers, young adults, and even older adults.

OUR CHANGING UNDERSTANDING OF ADOPTION

In doing some research on the history of adoption, I learned that our ideas about adoption started after World War II, when the U.S. took the lead in promoting both domestic and international adoptions.[5] In 1950 it was a new idea that one could bring the baby of a stranger of any class, race, or ethnicity from thousands of miles away, or even from halfway around the world, from China or India, into a private family and think that if you treated the baby as your biological child, his or her individual future would be determined completely by how you nurtured. It was also a new idea that adoption—a legal, contractual relationship which created a family of fictive kin—was now to be seen as no different from a biological family.

Previously, in numerous societies, including our own, children were often informally or legally adopted within the community, most often by relatives. In the U.S., before the 1920s, birth and adoptive parents often knew each other. Single women, especially unmarried professional women, were encouraged to adopt children, but by 1945 this was no longer permitted. Only in the 1940s were US adoption records legally closed and adoptees and adoptive parents barred from knowledge of, or contact with, birth parents. Only then were original birth certificates hidden and new ones drawn up listing the adoptive parents as if they were the birth parents. As a sociologist, I have noticed that this attempt to make an adopted family identical to a biological one coincided with the decline of extended family ties and the idealization of the nuclear family.

Starting in the 1980s, a new idea of adoption became mainstream: open adoption. Seeing that secrecy and closed adoption records had a negative effect on adoptees, psychologists and adoption experts now advocated for open records and for maintaining ongoing contact between birth parents, adoptees, and adoptive parents. By 2015, 95 percent of people adopting domestically had some degree of openness in their adoption,[6] ranging from sharing an occasional e-mail with a birth parent, to a few face-to-face meetings, to ongoing and frequent contact.[7] Only 20 to 30 percent of adoptees under age eighteen have face-to-face contact with a birth parent, and this may be only a short meeting once or twice a year.

A number of adoption studies have found that open adoptions with higher levels of contact have positive effects not just on adoptees, but also on both biological and adoptive parents. All parties have more satisfaction with the adoption.[8] Perhaps this is true because the relationships between adoptees, birth parents, and adoptive parents—relationships which are always present psychologically— are now out in the open.[9] Most studies of open adoption, however, do not consider how adoptive parents can benefit from greater knowledge of their adopted child's background, including information on personality and cognitive traits as well as family histories of substance abuse, mental illness, and learning disabilities.[10]

My work contributes to adoption theory and practice by adding to our understanding of the benefits of open adoption. Open adoption helps adoptive parents learn more about an adopted child's heritage, permitting them to try to mold the environment to the abilities and needs of their child. Behavioral genetics indicates that the more we know about birth families, the better parents we can be. Thomas Bouchard, a prominent behavioral geneticist, concluded that parents and professionals can have an impact on a child's highly heritable traits, but such interventions will be most effective when tailored to each specific child's talents and inclinations.[11] Beth Hall, the director of PACT: An Adoption Alliance, in Oakland, California, provided me with an example of how knowledge about birth family traits can help adoptive parents:

> I remember a case once where a kid who was kind of a tinkerer was placed with a family that was very goal-oriented. The family felt that he was not motivated and not "trying" in school, but when we talked with his birth family, it turns out that many of them were engineers and web designers who had similar histories in school but later bloomed into professional tinkerers, if you will, who learned by doing. Once the adoptive parents were enlightened about this, they were able to take a different attitude and actually promote his tinkering. They stopped taking down his "projects" because it was "time to clean up."

IMPLICATIONS FOR ADOPTION PRACTICE

Based on my research, I'd give the following advice to adoptive parents:

- No matter how excellent your parenting skills, you as an adoptive parent cannot predict or control the adult outcome of your adopted son or daughter. Their adult life will depend on a complex mix of genetic inheritance, along with the home, school, and community environment you provide for them and the choices they make within that environment—choices that will be influenced by their genetic inheritance. Prenatal and postnatal environments over which you had no control may also be important.

• Because of these processes, your adult adopted son or daughter may very well be unlike you in personality, interests, and achievements, even when strong and loving bonds persist into adulthood. Your adopted daughter or son will in all probability be quite different from your biological children or children you adopt from more than one family. When a child—even if adopted as an infant—has a personality and interests different than those of the adoptive parents, adoption can easily lead to low self-esteem, especially if the parents have high achievement expectations for their child. If the child does not or cannot meet these expectations, she or he can feel guilty and inadequate. This is a more probable explanation for so-called underachievement in adoptees than some sense of psychological loss supposedly inherent in adoption.

• In years past, adoptive parents sometimes preferred to adopt internationally in order to avoid any relationship with birth families. From my perspective, which emphasizes the importance of genetics in human development, international adoption without access to birth families poses a disadvantage for the adoptive parent and the adoptee. Recognizing this, some adoptive parents have tried to find foreign birth families when their adoptive child is small.[12] When there is no contact with birth parents, adoptive parents must work harder to discover their child's interests, abilities, and liabilities.

My research also has implications for the rights and responsibilities of birth parents. They place a child with another family, but the child carries their genes and will inherit some of their traits and those of their extended families. Through agreeing to (and sometimes fighting for) ongoing contact with the child, birth parents can contribute to the adoptee's understanding and acceptance of self. Birth parents and their extended birth families can provide adoptive parents with information on extended family characteristics, information which will help them understand the child they are raising and provide better parenting.

As for me, I wish I'd had the courage to open my son's adoption sooner. If, during his teenage years, I had known about his birth parents' substance abuse, I would have been less anxious and confused, could have sought effective help, and would have taken a stronger stand against drugs and alcohol. This may or may not have made a difference for Marco, but meeting his half-brother at sixteen rather than at twenty-six could have been decisive. In all likelihood, his brother would have told him ten years earlier what it was like growing up with addicted parents, and how it motivated him to never take a drink or use a drug.[13] Marco, however, feels it would have been hard for him as a teenager to find out that his birth father was then in jail and that his birth mother was poor and had addiction problems. If, through open adoption, this information had been there from the beginning, perhaps there could have been gradual acknowledgment and acceptance.

Had I then had the knowledge I have now, I would have accompanied Marco when he was younger for a reunion with his birth parents, making it less likely that he could have related to them through substance use.[14] I wish too that at an earlier age, Marco could have known his birth uncle, who, like me and like my father, has a Ph.D. He might have believed more in his own intelligence. Whether he chose to participate or not, Marco could have seen that some of his birth family are also in our "family business" of education. My wish, however, demonstrates how hard it is for adoptive parents to give up the desire to have a child who shares some of their own characteristics and interests.

Behavior geneticists have started to indicate how their research could be used in adoption counseling. For example, David Reiss and Leslie Leve discuss how knowledge about birth parents' temperament and psychological characteristics, knowledge gained without genetic testing, helps predict a toddler's or an adolescent's behavior. With this kind of information, adoptive parents can be educated and counseled on how to react so as to maximize positive gene expression (as seen in the tinkering example above) and minimize the effects of a genetic predisposition for substance abuse or mental illness. Reiss and Leve conclude that helping parents respond appropriately to adverse behavior from their children might accomplish two ends: "improve the quality of parent-child relationships and moderate the probability that adverse genetic influences are expressed fully, if at all, in the patterns of the child's adjustment."[15]

Too often, cultural images still present nature (or genes) as destiny leading to a policy of fatalism. In contrast, studying behavioral genetics made me see that genes and environment interact and both are subject to change. This position promotes the idea of a *personalized social policy*, similar to the idea of personalized medicine. Sociologist Aaron Panofsky concludes that personalized social policy could use behavior genetic studies "to enable policy makers and social service providers to craft differential interventions most likely to be effective for particular sets of people in particular circumstances."[16] I think this idea is especially relevant to adoption.

All of the complex questions which adoptive parents must now face indicate a need for adoption agencies to provide more postadoption education and support for adoptive parents, not just when the adoptees are children or teenagers, but also when some are troubled adults. Behavioral-genetics research can be incorporated into such practice. Richard Barth, dean of the University of Maryland School of Social Work, a widely published researcher about adoption and himself an adoptive parent, sums up the work adoption professionals face in supporting adoptive parents: "There is no certainty in family life, but providing families with reasonable expectations can help them enter adoptions with a sense of choice, purpose, and no less enthusiasm about providing love and care for a child who is likely to be more different than similar to children they already have and to the children they once were."[17]

My story argues for the importance of incorporating the findings of behavioral genetics into adoption practice, but my experience with transracial adoption has been a complicating factor, indicating how my idealistic naiveté about race and social class has been challenged. I've learned that parental monitoring of the environment—neighborhood, school, peers—has some impact on the adoptive child's development. But in my case, what the adopted child needed sometimes conflicted with my needs. This may be especially true for single, gay, lesbian or transgender parents, and/or if one's child is of a different race. A nontraditional parent, especially one that is not part of an extended family, needs a community to help with parenting, but what helps the parent may not be an ideal environment for the adoptee. Adoption agencies and adoption education should help adoptive parents figure out how much they can and are willing to change basic social choices, such as their geographical location, friends, religious affiliation, and race and class connections, in order to find the best fit for their adoptive children. Adoptive parents must then ask themselves how effective a parent they can be if they change their environment to one in which they are not at home. They need to ask how they can compromise between their needs and those of their adopted child.

IMPLICATIONS FOR ADOPTION RESEARCH

Once I discovered that behavioral genetics research was based on the study of adoptive families, I asked myself why adoption researchers had remained ignorant of, or uninterested in, behavioral genetics. After all, both behavioral genetics and adoption studies originated in psychology. The quantitative methods used in behavioral genetics are standard in social science and do not entail any detailed knowledge of genetics.

Part of the answer is that behavioral geneticists were uninterested in adoption. Their main focus was on obtaining legitimacy as a science. They sought to distance themselves from past disputes about IQ and race and they were not interested in any new policy debates. For them, adoption merely provided the possibility of a natural experiment, free of any manipulation of human subjects.[18] In an intact family, if a depressed mother had a depressed child, researchers could not determine how much such depression was due to genetic transmission and how much to the effects of the family environment. In an adopted family, with no contact with the birth families, the impact of the genes of the birth parents could be separated from nurturance in the new family.

But why were adoption researchers and policy makers, many of whom were also psychologists, uninterested in behavioral genetics? They might have been put off by the narrowness and complexities of the quantitative statistics used in behavioral genetics and by the field's tainted past. I believe a better explanation, however, is that adoption researchers were wedded to the dominant paradigm of psychological loss and to the empirical study of open adoption.

Those theorists most noted for the theory of psychological loss through adoption, especially David Brodzinsky and Harold Grotevant, recommended reunions with birth families to heal the loss, or better yet they wanted open adoption to maintain ties between birth and adoptive families from the beginning in order to prevent it. Starting in the late 1980s, advocacy of open adoption, with varying degrees of connection, became the major policy change in adoption theory and practice. Since behavioral genetics findings reinforce the possible benefits of open adoption, I thought they should have been interested. I decided to contact a few of the leading adoption researchers whom I had met at conferences to see if they had any interest in, or connection to, the field of behavioral genetics.

I e-mailed David M. Brodzinsky, a well-known psychologist who is co-author of *Being Adopted: The Lifelong Search for Self* (1993) and coeditor of two important anthologies, *The Psychology of Adoption* (1990) and *Psychological Issues in Adoption* (2005), along with being the author of articles about the complexities of open adoption. He replied, "I haven't read this literature for some time." I was surprised since recently I had heard him give a talk at an adoption conference where his PowerPoint listed genetics as the first factor in "Development Issues Related to Adoption Adjustment." He said: "Most of the characteristics associated with children's physical, psychological, social and academic adjustment are at least moderately influenced by heredity. [This gives us] reason to believe that adopted children are at increased risk for certain types of biologically based problems." However, for whatever reason, Brodzinsky has not incorporated the findings of behavioral genetics into his own research agenda.

I also contacted Harold Grotevant, professor of psychology and director of the Rudd Adoption Research Program at the University of Massachusetts, Amherst, and the author of many papers about the impact of open adoption. He told me that as a graduate student at the University of Minnesota in the 1970s, he had worked on one of the well-known behavioral genetics studies using adoptees. But his research interests had veered off in other directions, to the study of open adoption. While not hostile to behavioral genetics, Grotevant didn't see much need to integrate their findings into adoption studies.

A younger adoption advocate, Adam Pertman, the author of *Adoption Nation* (2011) and at the time executive director of the Donaldson Adoption Research Institute in New York City, seemed unfamiliar with the term *behavioral genetics* and asked a research assistant to respond to me. *Adoption Nation* does include the theory of psychological loss in adoption but nothing on genetics, demonstrating how in this field, as in many others, a dominant theoretical paradigm can eliminate other perspectives.

I advocate that those engaged in adoption research should familiarize themselves with behavioral genetics research, especially with the data gathered from adoptive families, birth families, and adoptees. I have cited only a small proportion of behavioral genetics studies that use adoptive families as subjects, particularly those focused on personality, intelligence, and addiction. There are a lot more out there. Behavior geneticists themselves have just begun to interpret their findings for the adoption community.[19]

Ironically, there may not be a lot of new behavioral genetic studies using adoption data, for open adoption is undermining the separation of genes and environment that previously attracted behavioral geneticists. Sibling studies, however, will be enriched as behavioral geneticists compare adoptees with their half and full siblings who stayed with the birth mother, and with other adopted and biological children in their adoptive families. Twin studies, whose findings are comparable to those from adoption studies, will continue to uncover new findings in behavioral genetics that will also shed light on adoption.[20]

ETHICAL CONCERNS

Emphasizing the importance of genetics in adoption raises new ethical questions. What if prospective parents were to ask for testing of birth parents or of the child before they decide whether to adopt? At least for the foreseeable future, this is not a valid concern, since the attempt to link behavior with specific genes has not been successful.[21] Rather, behaviors are now seen as the result of multiple genes, each with small effects that can be altered by each other and by the environment. If adoptive parents want to get more information about a birth family's heredity, the most effective way is through more interaction with them in open adoption. The perspective presented here also puts more responsibility on adoption agencies to get as much information as possible about birth family histories and to fully disclose this information to adoptive parents.

The American Society of Human Genetics and the American College of Medical Genetics have already established specific guidelines for adoption. In a joint statement, "Genetic Testing in Adoption," they affirm that tests on children up for adoption should consist of only those tests done on all babies. Genetic tests on any infant are recommended only for abnormalities that exist in childhood for which intervention is possible. Infants should not be tested for diseases for which there is a genetic marker but which appear only in adults, for example, Huntington's Disease. Instead, testing should wait until the individual is old enough to make a decision on whether to be tested. The organization's statement concludes that no test should be used to detect genetic variations within the normal range: "Allowing adoptive parents to gain access to a child's predisposition to a genetic condition that may never develop treats the adopted child differently from other children of a similar age and places a burden of perfection on the adopted child. Children do not come with guarantees."[22]

A NEW KIND OF FAMILY

Perhaps the most innovative implication of this book is my advocacy for forming extended family ties incorporating both biological and adoptive family members. In this model, adoptees, adoptive parents, and birth parents all have relationships to each other and to relatives beyond the nuclear family.

In the past, adopting a child has meant only incorporating an individual child into one's own nuclear and extended family. Now prospective adoptive parents need to consider that open adoption could mean extending family ties to include birth parents and some of their families. Today, a few families are forging such new kinship networks between adoptive and biological families from the time that the adopted child is an infant or toddler. Although such innovation is often difficult, especially when there are class differences or family dysfunctions involved, we are beginning to hear their voices.[23] Forming such extended families may provide tremendous benefits to adoptees and birth parents, but also may lead to more satisfied and effective adoptive parents, ones with more knowledge and fewer illusions.

My experience and that of my son document that an adult reunion can also lead to extended family ties. Here, the adoptee has more say in creating the extended family. Marco chose his Creole half-sister and I chose his black stepmother with whom to have ongoing ties within his birth families. Both these relatives helped make the relationships happen. Even in adult adoption searches, where an adoptee finds a birth parent who is deceased or doesn't want a reunion, she or he may find siblings, aunts, uncles, or grandparents with whom to relate. Or maybe his adoptive parents don't want to extend family ties but the adult adoptee may have siblings or cousins within the adoptive family that will want to relate to some of those in the birth family.

In a society where family life is being drastically transformed,[24] and where genetic ties are given increasing importance through the wide availability of new reproductive technologies, through inexpensive genetic testing, and through the possibility of internet connection with previously unknown biological kin,[25] theory and research on adoption are now mainstream. I hope that my story and my research will contribute to their continued productivity.

ACKNOWLEDGMENTS

I owe the most thanks to two fellow writers and friends, Gayle Greene and Mardi Louisell. In several different writing groups, and individually, they read countless drafts and provided general suggestions and detailed editing. I learned from them and became a better writer.

I met Andrew Solomon when he read in San Francisco in November 2016. I had admired both *The Noonday Demon* and *Far from the Tree*, and I now found him open and approachable. In talking with him, I found that we had a common interest in nontraditional families, and adoption was one of the topics he planned to write about. Because of that interest he agreed to read my manuscript. I was pleased with his enthusiastic response and his desire to interview me and Marco (separately) for his Amazon audio book, *New Family Values,* and his book *Who Rocks the Cradle,* forthcoming from Scribners. His unwavering support for the value of my story and research sustained me over the difficult course of revision and of finding a publisher.

I want to thank the many colleagues and friends who read all or part of earlier drafts of the manuscript and offered valuable comments. They are (in alphabetical order) Ann Berlak, Wini Breines, Laura Callen, Barbara Epstein, Marilyn Fabe, LeAnn Fields, Tina Gillis, Judy Grether, Beth Hall, Elsa Johnson, Katherine Johnson, Claire Kahane, Naomi Katz, Marlene Legates, Nicole Magnuson, Wendy Martin, Glenna Matthews, Randy Milden, Meredith Minkler, Judy Newton, Ilene Philipson, Mary Pittman, Bob Prentice, Jill Stanton, Gail Steinberg, Gretchen Strain, Diane Trimberger, Dorothy Wall, and Alice Wexler.

I'm grateful too to those who have heard me talk (publicly, in a support group, or privately) about the book. Some offered technical help, and all provided encouragement and support. In alphabetical order, they are Martin Bennett, Susie Bluestone, Michael Burowoy, Nancy Carleton, Myrna Cozen, Lindsey Crittenden, Bella De Paulo, Ellen DuBois, Clare Fischer, Victor Garlin, Robert Girling, David Glotzer, Sue Gretner, Sallie Hanna-Rhyne, Barbara Heyns, Lynne Hollander, Ramsey Kanaan, Sherry Keith, Jack Lasche, Rosemary Trimberger Lasche, Marge Lasky, Kathleen Lilley, Sasha Lilley, Ellen Meeropol, Charlotte Meyer, Rachel Moran, Veronica Moreno, Margaret Nelson, Chris Olsen, Patricia Ramsey, Claudia Robbins, and Jim Stockinger.

This book has benefited from the excellent, efficient, and supportive editorial and marketing staff at Louisiana State University Press. I especially want to thank Margaret Lovecraft, but also Alisa Plant, Kate Barton, Catherine Kadair, Mandy Scallan, LB Kovac, and James Wilson. I also thank my freelance editor Jo Ann Kiser and indexer Scott Smiley.

Lastly, I thank my son, Marco, who has enriched my life. His empathy, openness, intelligence, and ability to change made this book possible.

NOTES

CHAPTER 1
HOW IT ALL BEGAN

1. Four of my friends in the late 1970s—all sociologists and feminist scholars—wrote influential critiques of an older feminist sociologist who wanted to incorporate a biological perspective in feminist analysis of sex, gender, and the family. At the time, I sided with my friends. See Margaret Cerullo, Judith Stacey, and Wini Breines, "Alice Rossi's Sociobiology and Anti-Feminist Backlash," and Nancy Chodorow, "Considerations on a Biosocial Perspective on Parenting," both in the *Berkeley Journal of Sociology* 22 (1977–78).

2. A study in the 1980s of attitudes toward single mothers found that those single women who adopted were viewed more positively than those who gave birth. Ruth Mechaneck, Elisabeth Klein, and Judith Kupperssmith, "Single Mothers by Choice: A Family Alternative," in *Women, Power and Therapy,* ed. Marjorie Braude (New York: Harrington Press, 1988), 266.

3. In the 1960s, birth mothers gave up 80 percent of white babies born out of wedlock. In 1980, mothers released only 4 percent of such babies. Ellen Herman, *Kinship by Design: A History of Adoption in the Modern United States* (Chicago: University of Chicago Press, 2008), 255.

4. See Sandra Patton, *Birth Marks: Tranracial Adoption in Contemporary America* (New York: New York University Press, 2000), 48–50.

5. See Randall Kennedy, *Interracial Intimacies: Sex, Marriage, Identity, and Adoption* (New York: Pantheon Books, 2003), chap. 10.

6. Others writing about adoption have used the imagery of ghosts. See *A Ghost at Heart's Edge: Stories and Poems of Adoption,* ed. Susan Ito and Tina Cervin (Berkeley: North Atlantic Books, 1999). Psychologist David Brodzinsky says the "the lost birth parents often linger as *ghosts* in the mental and emotional life of adopted persons." See "A Stress and Coping Model of Adoption Adjustment," in *The Psychology of Adoption,* ed. David Brodzinsky and Marshall Schechter (New York: Oxford University Press, 1990), 9.

7. Some years later, back in Berkeley, I saw at the Berkeley Repertory Theater two short plays about Louisiana by the playwright Anne Galjour, originally from southern Louisiana. In the program, along with other educational material, was a recipe for turtle sauce piquante, which called for 2–3 lbs. of turtle meat, or rabbit, frog legs or squirrel.

8. For an excellent presentation of Louisiana Creole culture and heritage, see Elista Istre, *Creoles of South Louisiana: Three Centuries Strong* (Lafayette: University of Louisiana at Lafayette Press, 2018).

9. Bliss Broyard, *One Drop: My Father's Hidden Life—A Story of Family Secrets* (New York: Little Brown, 2007), 298.

10. Annie Murphy Paul, *Origins: How the Nine Months before Birth Shape the Rest of Our Lives* (New York: Free Press, 2010), 22.

CHAPTER 2
SEARCHES

1. For an alternative view, now called "open adoption," see Joyce M. Pavao, *The Family of Adoption*, rev. ed. (Boston: Beacon Press, 2005).

2. Ulrich Muller and Barbara Perry, "Adopted Persons' Search for and Contact with Their Birth Parents II: Adoptee-Birth Parent Contact," *Adoption Quarterly*, 4, no. 3 (2001): 39–62.

3. In her interviews with adult transracial adoptees, Sandra Patton found that "the people who longed to know their birth fathers were generally biracial adoptees with a white birth mother and a black birth father." See her *Birth Marks: Transracial Adoption in Contemporary America* (New York: New York University Press, 2000), 115.

4. Seventy-five percent of birth parents reacted positively when found by an adult adopted person (Patton, *Birth Marks*, 43).

5. Sandra Patton in her interviews with adult transracial adoptees found that an "ability to interact with different groups and cultures with ease emerged repeatedly in my interviews." She concluded that this ability was a major survival skill of transracial adoptees that they used to navigate through their complex lives (Patton, *Birth Marks*, 10).

6. See David Brodzinsky, Marshall Schechter, and Robin Morantz, *Being Adopted: The Lifelong Search for Self* (New York: Anchor Books, 1993).

7. See Rickie Solinger, *Wake Up Little Susie: Single Pregnancy and Race before Roe versus Wade*, 2nd ed. (New York: Routledge, 2000).

8. In writing this book, I have confronted some of the class and race biases I had at the time and have looked at how experience proved them to be untrue. Although it is personally embarrassing, I believe that avoiding denial and bringing such prejudices out in the open is a step toward their eradication, personally and socially.

9. Louisiana imprisons more of its people than any other state in the US and therefore than any other place in the world. See Charles Blow, "Plantations, Prisons and Profits," opinion piece in the *New York Times*, May 26, 2012.

10. See Steven Pinker, *The Blank Slate: The Modern Denial of Human Nature* (New York: Penguin Books, 2002).

11. A major center for behavioral genetics research is the Colorado Adoption Project

(CAP) established in 1976 by the Institute of Behavioral Genetics at the University of Colorado. Their longitudinal study of more than 2,400 individuals—birth parents, adoptive parents, adoptees and siblings—is ongoing, with nearly 90 percent of the sample still participating.

12. For a good textbook on behavioral genetics, see Robert Plomin, John C. DeFries, Valerie Knopik, and Jenae Neiderhiser, *Behavioral Genetics,* 6th ed. (New York: Worth Publishers, 2013).

13. See Richard C. Francis, *Epigenetics: The Untimate Mystery of Inheritance* (New York: W. W. Norton, 2011).

14. While accepting the importance of genetic inheritance, both behavioral geneticists and outside critics discuss whether heritability estimates are too high or too low, and whether they have any meaning at all. They argue about whether you can really separate the impacts of heredity and environment. See Aaron Panofsky, *Misbehaving Science: Controversy and the Development of Behavior Genetics* (Chicago: University of Chicago Press, 2014); Erik Parens, Audrey R. Chapman, and Nancy Press, eds., *Wrestling with Behavioral Genetics: Science, Ethics, and Public Conversation* (Baltimore: Johns Hopkins University Press, 2006).

15. See Thomas J. Bouchard and Matt McGue, "Genetic and Environmental Influences on Human Psychological Differences," *Journal of Neurobiology* 54, nos. 1–3 (2003): 4–45.

16. I was surprised to find that few researchers in adoption studies have looked at this data. Two adoption researchers who have incorporated some of the findings of behavioral genetics are Richard Barth, "Outcomes of Adoption and What They Tell Us about Designing Adoption Services," *Adoption Quarterly* 6, no. 1 (2002): 45–60; and David Howe, *Patterns of Adoption: Nature, Nurture, and Psychosocial Development* (Oxford: Blackwell Science, 1998).

17. Ninety percent of the biological parents and 95 percent of adoptive parents were white—with the rest Hispanic and Asian American—and they were from roughly equal social classes. Robert Plomin, David W. Fulker, Robin Corley, and John C. DeFries, "Nature, Nurture and Cognitive Development from 1 to 16 Years: A Parent-Offspring Adoption Study," *Psychological Science* 8, no. 6 (November 1977): 442–47.

18. S. J. Wadsworth, R. P. Corley, J. K. Hewitt, R. Plomin, and J. C. DeFries, "Parent-Offspring Resemblance for Reading Performance at 7, 12 and 16 Years of Age in the Colorado Adoption Project," *Journal of Child Psychology and Psychiatry* 43, no. 6 (2002): 769–74.

19. Megan M. McClelland, Alan C. Acock, Andrea Piccinin, Sally Ann Rhea, and Michael C. Stallings, "Relations between Preschool Attention Span-Persistence and Age 25 Educational Outcome," *Early Childhood Research Quarterly* 28 (2013): 314–24.

20. Psychologists sometimes conceptualize and name these traits somewhat differently.

21. John C. Loehlin, Joseph M. Horn and Lee Willerman, "Hereditary, Environment and Personality Change: Evidence from the Texas Adoption Project," *Journal of Personality* 58, no. 1 (1990): 221–43, p. 231.

22. The subjects in this study were 229 adopted children and 83 biological children of

the adopted parents. John C. Loehlin, Lee Willerman, and Joseph M. Horn, "Personality Resemblance in Adoptive Families: A 10-Year Followup," *Journal of Personality and Social Psychology* 53, no. 5 (1987): 961–69.

23. Ibid., 964.

24. Ibid., 968.

25. John Loehlin, Joseph M. Horn, and Jody L. Ernst, "Genetic and Environmental Influences on Adult Life Outcomes: Evidence from the Texas Adoption Project," *Behavior Genetics* 37 (2007): 463–76.

26. Behavioral genetics has done a lot of work on the genetics of personality disorders, especially schizophrenia, which I will not consider in this book. See Robert Plomin, John C. DeFries, Valerie Knopik, and Jenae Neiderhiser, *Behavioral Genetics*, 6th ed. (New York: Worth Publishers, 2013), chaps. 14–17.

CHAPTER 3
TARNISHED JEWELS

1. I learned at a conference of the American Adoption Congress that such sexual attraction between reunited birth parents is not unusual.

2. There are many memoirs by adult adoptees about their reunions with birth parents. A few I like are: Jackie Kay, *Red Dust Road* (New York: Atlas, 2010); A. M. Homes, *The Mistress's Daughter* (New York: Penguin Books, 2007); Jane Jeong Trenka, *The Language of Blood* (Minneapolis: Graywolf Press, 2003); Sarah Saffian, *Ithaka: A Daughter's Memoir of Being Found* (New York: Basic Books, 1998); and Laura Dennis, *Adoption Reunion in the Social Media Age: An Anthology* (Redondo Beach, CA: Entourage Publishing, 2014).

3. Robert Plomin et al., *Behavioral Genetics*, 6th ed. (New York: Worth Publishers, 2013), 145.

4. Robert L. DuPont, M.D., *The Selfish Brain: Learning from Addiction* (Center City, MN: Hazelden, 2000), xxiv.

5. Sheff quoted the 2013 edition of the DSM (*The Diagnostic and Statistical Manual of Mental Disorders*), which is the official publication of the American Psychiatric Association used by doctors and mental health professionals to diagnose specific mental illnesses. This DSM eliminates drug abuse as separate from addiction. See David Sheff, *Clean: Overcoming Addiction and Ending America's Greatest Tragedy* (New York: Houghton Mifflin, 2013), 78–79.

6. Ibid., 98.

7. Ibid., 10–11.

8. Ming D. Li and Margit Burmeister, "New Insights into the Genetics of Addiction," *Nature Reviews/Genetics* 10 (April 2009): 225.

9. Matt McGue, Steve Malone, Margaret Keyes, and William G. Iacono, "Parent-Offspring Similarity for Drinking: A Longitudinal Adoption Study," *Behavior Genetics* 44 (2014): 620–28.

10. All families were from Minnesota and the adoptees were placed prior to two years of age, with 96 percent placed prior to one year. Ninety-five percent of adopted parents, control parents, and control children were white, but the majority of adoptees were non-white, with 67 percent being Korean and 12 percent from other ethnic and racial backgrounds.

11. Remi J. Cadoret, M.D., et al., "Adoption Study Demonstrating Two Genetic Pathways to Drug Abuse," *Archives of General Psychiatry* 52 (January 1995): 52.

12. The number of subjects was small—thirty-three adoptees with substance abuse birth parents and thirty-nine adoptees from "clean" biological families.

13. This study did not look at environmental factors outside the home, which other researchers on the initiation of substance abuse see as important. I will examine this research in the next chapter.

14. Susan E. Young et al., "Genetic and Environmental Vulnerabilities Underlying Adolescent Substance Use and Problem Use: General or Specific," *Behavior Genetics* 36, no. 4 (2006): 613.

15. Eric J. Nestler, "Is There a Common Molecular Pathway for Addiction?" *Nature Neuroscience* 8, no. 11 (2005): 1445–1449.

16. Kenneth Kendler and Carol A. Prescott, *Genes, Environment, and Psychopathology: Understanding the Cause of Psychiatric and Substance Use Disorders* (New York: Guilford Press), 101.

17. The Swedish study, published in 2012, was based on data obtained from comprehensive national registries for more than 18,000 adopted children born between 1950 and 1993. Sweden has a unique individual personal identification number that is assigned at birth and is used by all Swedish residents for their lifetime. By assigning another number for each I.D. (to maintain confidentiality), the researchers were able to obtain indications of drug abuse for birth parents, adoptive parents and adult adoptees from a number of national registries for health, prescription drugs and crime. The adoptees were an average age of forty-six years when the data was collected.

18. Kenneth Kendler, M.D., et al., "Genetic and Familial Environmental Influences on the Risk for Drug Abuse: A National Swedish Adoption Study," *Archives of General Psychiatry* 69, no. 7 (2012): 690–97.

19. Kenneth Kendler et al., "A Multivariate Twin Study of the DSM-IV Criteria for Personality Disorder," *Biological Psychiatry* 71 (2012): 247–53.

20, S. Alexandra Burt and Jenae Neiderhiser, "Aggressive versus Nonaggressive Anti-social Behavior: Distinctive Etiological Moderation by Age," *Developmental Psychology* 45, no. 4 (2009): 1164–1176.

21. This report compared the results of sixty-four studies (only a few by behavioral geneticists). See James MacKillop et al., "Delayed Reward Discounting and Addictive Behavior: A Meta-analysis," *Psychopharmacology* (2011): 216, 305–21.

22. Ulrich Muller and Barbara Perry, "Adopted Persons' Search for and Contact with Their Birth Parents I: Who Searches and Why?" *Adoption Quarterly* 4, no. 3 (2001): 5–37.

1. Berkeley, with a population of 110, 500, is contained in a small area, with 6,362 persons per square mile. Compare this to Oakland (with 5,009 people per square mile) and to the state and national average (239 and 87 persons per square mile, respectively).

2. Victoria Pardini, "Black Population Declines by 20 Percent over Past Decade," *Daily California,* April 14, 2011.

3. Oakland, in contrast, is close to the state and national average of 27 percent of adults who attain this educational level.

4. Aaron Glantz, "Gap between Rich and Poor in Area Is Widest in Berkeley," *San Francisco Chronicle,* November 19, 2011.

5. This left was *new* in rejecting bureaucratic structures of universities, parties, and states, but also in seeking to create new communal forms for work and family.

6. The historian Timothy Miller in his 1999 book, *The 60s Communes: Hippies and Beyond,* says that 200–800 communes existed in Berkeley in 1972. Eric Raimy in his 1979 book, *Shared Houses, Shared Lives,* quotes the *San Francisco Chronicle* in 1978 as saying that there were almost six thousand communal households in Berkeley (New York: Houghton Mifflin). Whatever the number, my household was not unique.

7. Timothy Miller, *The 60s Communes: Hippies and Beyond* (Syracuse, NY: Syracuse University Press, 1999), chap. 8.

8. Amitai Etzioni, *The New Golden Rule: Community and Morality in a Democratic Society* (New York: Basic Books, 1996), 127.

9. Benjamin D. Zablocki, "What Can the Study of Communities Teach Us about Community?" in *Autonomy and Order: A Communitarian Anthology,* ed. Edward W. Lehman (New York: Rowman & Littlefield, 2000), 79.

10. The continuation of individualism within the communal movements of the 1960s and 1970s is documented in *West of Eden: Communes and Utopia in Northern California,* ed. Iain Boal, Janferie Stone, Michael Watts, and Cal Winslow (Oakland, CA: PM Press, 2012).

11. Rosanna Hertz, *Single by Chance, Mothers by Choice: How Women Are Choosing Parenthood without Marriage and Creating the New American Family* (New York: Oxford University Press, 2006), 61–62.

12. Susan Golombok looks at the evidence in *Modern Families: Parents and Children in New Family Forms* (New York: Cambridge University Press, 2015), 196.

13. Bella DePaulo, *How We Live Now: Redefining Home and Family in the 21st Century* (New York: Atria Books, 2015), chap. 5, "Not-So-Single Parents."

14. Thomas G. O'Connor et al., "Are Associations between Parental Divorce and Children's Adjustment Genetically Mediated? An Adoption Study," *Developmental Psychology* 36, no. 4 (2000).

15. The assumption was that those adoptees with a genetic predisposition to addiction

would be evenly distributed in the divorced and intact families. Because the children of divorce were such a small number (only fifteen children), this assumption might not be true. On measures of self-esteem, social competence, school achievement, and behavioral and emotional problems there were few significant differences between adoptees in the two types of families, and those that existed were very moderate. Twelve-year-olds in the control group of biological families also had significantly higher use of substances in divorced families, but it was not as high as in the adoptive families.

16. Remi J. Cadoret et al., "An Adoption Study of Genetic and Environmental Factors in Drug Abuse," *Archives of General Psychiatry* 43 (1986): 1131–1136.

17. S. Alexandra Burt et al., "Parental Divorce and Adolescent Delinquency: Ruling Out the Impact of Common Genes," *Developmental Psychology* 44, no. 6 (2008). They studied adolescents from 406 adoptive families and 204 biologically related families where all the families had at least two siblings.

18. For a summary of this research, see Michael Rutter, *Genes and Behavior, Nature-Nurture Interplay Explained* (Malden, MA: Blackwell Publishing, 2009), chaps. 5, 9, 10.

CHAPTER 5
CHILDHOOD TRAUMA

1. Recently, I talked about daycare with my niece and nephew, both of whom are married in families with two working parents. I was pleased to find out that their infants are in professional child-care centers. They can at any time log into cameras that monitor the centers and see their babies and their caregivers.

2. For an explanation and critique of this aspect of black child rearing, see Brittany Cooper, "The Racial Parenting Divide," http://www.salon.com/2004/09/16.

3. See Carolyn S. Schroeder and Betty N. Gordon, *Assessment and Treatment of Childhood Problems* (New York: Guilford Press, 2002), chap. 7, "Sexuality and Sexual Problems," 245.

4. In a comprehensive quantitative sociological study of sexuality in the US, adult male informants who had sexual contact with adults as children reported that 40 percent of the contacts were with a family friend and only 4 percent with a teacher. Edward O. Laumann, John H. Gagnon, Robert T. Michael, and Stuart Michaels, *The Social Organization of Sexuality: Sexual Practices in the United States* (Chicago: University of Chicago Press, 1994), 373.

5. Schroeder and Gordon, *Assessment and Treatment*, contend that "maternal support, especially at the time of disclosure, has consistently been found to be related to a child's positive adjustment" (235).

6. Research indicates that children are more at risk of abuse if they live in a household with a male who is not their genetic father. Lawrence Gagong and Marilyn Coleman, *Stepfamily Relationships: Development, Dynamics and Intervention* (New York: Kluwer Academic, Plenum Publishers, 2004), 155.

7. See Susan A. Clancy, *The Trauma Myth: The Truth about Sexual Abuse of Children and Its Aftermath* (New York: Basic Books, 2009), 71.

8. Academic researchers have relied on contemporary reports of parents and pre-school teachers. Although not ideal, such data, especially when the numbers are large, is better than that obtained from retrospective reports of college students and other adults. Some researchers also have found innovative ways to talk with children about sexuality. A study in 2000 interviewed (with parental permission) 147 children (84 boys and 63 girls) between the ages of two and six in preschools in three different German cities. None of the children up to age five demonstrated any knowledge of explicit sexual behavior of adults, and only a few five- and six-year-olds mentioned adult sexual activity. Renate Volbert, "Sexual Knowledge of Preschool Children," in *Childhood Sexuality: Normal Sexual Behavior and Development*, ed. Theor G. M. Sandfort and Jany Rademakers (New York: Haworth Press, 2000). Copublished simultaneously in *Journal of Psychology and Human Sexuality*, 12, nos. 1, 2 (2000).

9. More than 1,100 mothers with a child between ages two and twelve were recruited from pediatric clinics in Minnesota and Los Angeles. Mothers who suspected or could substantiate sexual abuse of their child were excluded. Mothers in the study filled out a long questionnaire about whether they had observed a large number of very specific sex acts and sexual touching. William N. Friedrich, Jennifer Fisher, Daniel Broughton, Margaret Houston, and Constance R. Shafran, "Normative Sexual Behavior in Children: A Contemporary Sample," *Pediatrics* 101, no. 4 (April 1998): 9.

10. In this age group, only 4.6 percent of boys touched another child's sex parts, 2.1 percent talked about sex acts, 1.4 percent undressed other children, and only 0.4 percent asked others to do sex acts or put objects in a vagina or rectum. The authors made clear that their sample could have included some parents who did not know that their children had been sexually abused.

11. Sandra K. Hewitt, *Assessing Allegations of Sexual Abuse in Preschool Children: Understanding Small Voices* (Thousand Oaks, CA.: Sage Publications, 1999), 8.

12. Ibid., 48, 76, 149.

13. For an extensive survey of this research, I relied on a book by Duke University psychologist Patricia J. Bauer, *Remembering the Times of Our Lives: Memory in Infancy and Beyond* (Mahwah, NJ: Lawrence Erlbaum Publishers, 2007).

14. Hewitt, *Assessing Allegations of Sexual Abuse*, 59–63.

15. Linda Meyer Williams, "Recall of Childhood Trauma: A Prospective Study of Women's Memories or Child Sexual Abuse," *Journal of Counseling and Clinical Psychology* 62, no. 6 (1994): 1167–1176.

16. Research on the impact of child sexual assault prevention programs is mixed about whether they facilitate disclosure during childhood. Some researchers have found an association between prevention programs and disclosure and other researchers have not found any association. One study found earlier disclosure as the result of exposure to prevention programs. This same study found that participation in a child sexual abuse prevention

program had no negative consequences for sexual satisfaction in adults. Laura E. Gibson and Harold Leitenberg, "Child Sexual Abuse Prevention Programs: Do They Decrease the Occurrence of Child Sexual Abuse?" *Child Abuse and Neglect* 24, no. 9 (2000): 1115–1125.

17. Studies of identical twins raised in the same family and sharing other risk factors found that when one twin was sexually abused as a child, she or he had a higher risk for adult psychopathology, especially substance abuse. See Kenneth S. Kendler and Carol A. Prescott, *Genes, Environment and Psychopathology: Understanding the Causes of Psychiatric and Substance Use Disorders* (New York: Guilford Press, 2006), 284.

18. Female adoptees (155) in Iowa who reported child sexual abuses were recruited for this study when they were young adults. (There were not enough males who admitted child sexual abuse to include them.) These women were compared with adopted women who had not experienced sexual abuse as children. The women who had been sexually abused as children were found to have more substance use, antisocial personality disorder, and depression, replicating the findings of other studies. The abused women then were divided into two groups: those who had one or both biological birth parents with a history of substance abuse, antisocial behavior, or depression compared to a second group whose biological parents did not have these problems. Controls were imposed for the behavior of adoptive parents. Steven R. H. Beach et al., "Impact of Child Sex Abuse on Adult Psychopathology: A Genetically and Epigenetically Informed Investigation," *Journal of Family Psychology* 27, no. 1 (2013): 3–11.

19. Richard C. Francis, *Epigenetics: The Ultimate Mystery of Inheritance* (New York: W. W. Norton, 2011).

20. One process of epigenetics that has been widely studied is methylation, where a methyl group (one carbon atom bonded to three hydrogen atoms) is attached to the gene. Ibid., 6.

21. Steven Beach et al., "Methylation of 5htt Mediates the Impact of Child Sex Abuse on Women's Antisocial Behavior: An Examination of the Iowa Adoptee Sample," *Psychosomatic Medicine* 73, no. 1 (2011): 83–87.

22. Eamon McCrory et al., "The Link between Child Abuse and Psychopathology: A Review of Neurobiological and Genetic Research," *Journal of the Royal Society of Medicine* 105, no. 4 (2012): 151–56.

CHAPTER 6

THE GOOD TIMES

1. See Rosanna Hertz, *Single by Chance, Mothers by Choice*, chap. 9.

2. Susan Golombok in *Modern Families* finds that statistics reporting more negative outcomes for children of single parents are "almost entirely accounted for by . . . economic hardship, maternal depression, and lack of social support, as well as factors that predate the transition to a single parent home, such as parental conflict" (196).

3. E. Kay Trimberger, *The New Single Woman* (Boston: Beacon Press, 2005), 192. The

single mothers studied by Rosanna Hertz in *Single by Chance, Mothers by Choice* had better experiences with Big Brothers (184–85).

4. Ann Arnett Ferguson, *Bad Boys: Public Schools in the Making of Black Masculinity* (Ann Arbor, University of Michigan Press, 2001).

5. Barrie Thorne found in her study of children at elementary schools that boys who were athletic and had good social skills could cross gender barriers with few negative consequences. *Gender Play: Girls and Boys in School* (New Brunswick, NJ: Rutgers University Press, 1993), 119, 123.

6. Trimberger, *The New Single Woman*, 189; Arlie Hochschild, *The Second Shift: Working Families and the Revolution at Home* (New York: Penguin Books, 1989).

7. Plomin et al., *Behavioral Genetics*, 6th ed., 110.

8. Timothy Biblarz and Adrian Raftery, "Family Structure, Educational Attainment and Socioeconomic Success: Rethinking the 'Pathology of Matriarchy,'" *American Journal of Sociology* 105, no. 2 (1999). Rosanna Hertz in her study of single mothers also found elementary schoolchildren suggesting men that their single mother could date (*Single by Chance, Mothers by Choice*, 188).

9. Erik Turkhimer, "Three Laws of Behavior Genetics and What They Mean," *Current Directions of Psychological Science*, 9, no. 5 (October 2000): 160.

10. Aaron Panofsky, *Misbehaving Science: Controversy and Development of Behavioral Genetics* (Chicago: University of Chicago Press, 2014), 142. Panofsky summarizes the work of Sandra Scarr, one of the leaders of behavioral genetics in the 1970s. For a more recent study making the same point, see Xiaojia Ge, Remi Cadoret, Rand Conger, and Jenae Neiderhiser, "The Developmental Interface between Nature and Nurture: A Mutual Influence Model of Child Antisocial Behavior and Parent Behaviors," *Developmental Psychology* 32, no. 4 (1996): 212.

11. Eleanor Maccoby, "Parenting and Its Effects on Children: On Reading and Misreading Behavior Genetics," *Annual Review of Psychology* 51 (2000): 1–27. p. 18.

12. This project recruited 338 triads (birth mother, adopted infant, and adoptive parents) from distinctive types of adoption agencies all over the US. In one-third of the triads, birth fathers were also included and intensively studied. Good information was available for most of the other birth fathers. Leslie Leve et al., "The Early Growth and Development Study: A Prospective Adoption Design," *Twin Research and Human Genetics* 1 (2007): 84–95 (PubMed 17539368).

13. Leslie Leve et al., "Structured Parenting of Toddlers at High versus Low Genetic Risk: Two Pathways to Child Problems," *Journal of the American Academy of Child and Adolescent Psychiatry* 48, no. 11 (2009): 1102–1109.

14. The adoptive mothers with their eighteen-month children were given a three-minute clean-up task. The interviewer asked each mother to have her child clean up a number of multipiece toys and put each toy set in a separate container. Only the child was to touch the toys, but mothers could provide instruction. Those mothers who provided structure spent more time giving instructions that asked for a specific action from the child, includ-

ing questions ("Where does this ring go?"), statements ("Let's put the duck in this box"), and directives ("Put all the cups in here").

15. Misaki Natsuaki et al., "Intergenerational Transmission of Risk for Social Inhibition: The Interplay between Parental Responsiveness and Genetic Influences," *Developmental Psychopathology* 25, no. 1 (2013): 261–78.

16. S. Alexandra Burt, "Rethinking Environment Contributions to Child and Adolescent Psychopathology: A Meta-Analysis of Shared Environmental Influences," *Psychological Bulletin* 135, no. 4 (2009): 608–34.

17. David Reiss et al., *The Relationship Code: Deciphering Genetic and Social Influences on Adolescent Development* (Cambridge, MA: Harvard University Press, 2000), 65.

18. Robert Plomin et al., *Behavioral Genetics*, 6th ed. (New York: Worth Publishers, 2013), 204.

19. Steven Pinker, *The Blank Slate: The Modern Denial of Human Nature* (New York: Penguin Books, 2004), 379.

20. Economist Bryan Caplan in his book *Selfish Reasons to Have More Kids: Why Being a Great Parent Is Less Work and More Fun Than You Think* (New York: Basic Books, 2011) uses behavioral genetics research on twins to conclude that the only thing a parent can control is whether your children appreciate you when they are adults because of the happy childhood you gave them.

CHAPTER 7
ADDICTION

1. Lisa N. Legrand et al., "Searching for Interactive Effects in the Etiology of Early-Onset Substance Use," *Behavior Genetics* 29, no. 6 (1999): 433–44.

2. Kenneth S. Kendler et al., "Genetic and Familial Environmental Influences on the Risk for Drug Abuse," *Archives of General Psychiatry* 69, no. 7 (2012): 690–97.

3. David Reiss and Leslie D. Leve, "Gene Expression outside the Skin: Clues to Mechanisms of Genotype x Environment Interactions," *Development and Psychopathology* 19, no. 4 (2007): 1009.

4. Beth Manke et al., "Genetic Contributions to Adolescents' Extrafamilial Social Interactions: Teachers, Best Friends and Peers," *Social Development* 4, no. 3 (1995): 238–56.

5. Robert Plomin et al., *Behavioral Genetics*, 6th ed. (New York: Worth Publishers, 2013), chap. 8.

6. Ibid., 253.

7. 2006 National Center on Addiction and Substance Abuse at Columbia University white paper, *Non-Medical Marijuana III: Rite of Passage or Russian Roulette?* chap. 12: "Marijuana Potency: Not Your Mother's Marijuana," 6.

8. George F. Koob and Michel Le Moal, *Neurobiology of Addiction* (Boston: Elsevier, 2006), 294.

9. "Marijuana Abuse," National Institute on Drug Abuse, Research Report Series, September, 2010.

10. Koob and Le Moal, *Neurobiology of Addiction,* 306.

11. One study found that only about one-third of regular marijuana users went on to develop abuse/dependency and that abuse was highly heritable. Ming T. Tsuang et al., "Genetic and Environmental Influences on Transitions in Drug Use," *Behavior Genetics* 29, no. 6 (1999): 473–79. Michael Rutter in a summary article, "Substance Use and Abuse: Causal Pathways Considerations," cites research that found that 99 percent of users of other illicit drugs had used cannabis first, but nearly two-thirds of marijuana users did not progress to other drugs. See *Child and Adolescent Psychiatry,* ed. Michael Rutter and Eric Taylor (Oxford: Blackwell Science, 2002), 459.

12. Carlton K. Erickson, *The Science of Addiction: From Neurobiology to Treatment* (New York: Norton, 2007), 25.

13. *The Health Effects of Cannabis and Cannabinoids: The Current State of Evidence and Recommendations for Research.* National Academies Press at www.nap.edu, 2017.

14. Arpana Agrawal and Michael T. Lynskey, "The Genetic Epidemiology of Cannibis Use, Abuse and Dependence," *Addiction* 101 (2006): 801–12.

15. K. Paige Harden et al., "Gene-Environment Correlation and Interaction in Peer Effects on Adolescent Alcohol and Tobacco Use," *Behavior Genetics* 38 (2008): 345.

16 Ibid., 10.

17. Danielle M. Dick et al., "Parental Monitoring Moderates the Importance of Genetic and Environmental Influences on Adolescent Smoking," *Journal of Abnormal Psychology* 116, no. 1 (2007): 213–18.

18. David Sheff, *Beautiful Boy: A Father's Journey through His Son's Addiction* (New York: Mariner Books, 2009), 52.

19. A behavioral genetics study that was done on toddlers seemed relevant to my experiences as a parent of a teen. Shannon T. Lipscomb and Leslie Leve et al., "Trajectories of Parenting and Child Negative Emotionality during Infancy and Toddlerhood: A Longitudinal Analysis," *Child Development* 82, no. 5 (2011): 1661–1675.

20. Xiaojia Gee et al., "The Developmental Interface between Nature and Nurture: A Mutual Influence Model of Child Antisocial Behavior and Parent Behaviors," *Developmental Psychology* 32, no. 4 (1996): 586.

21. Fragile Families and Child Wellbeing Study: Fact Sheet. https://fragilefamilies. princeton.edu/sites/fragilefamilies/files/ff_fact_sheet.pdf

22. Behavioral geneticists recognize that there is part of what they call the nonshared environment that cannot be accounted for in terms of conventional variables and which may be due to chance. See Eric Turkheimer, "Spinach and Ice Cream: Why Social Science Is So Difficult," in *Behavior Genetics Principles,* ed. Lisabeth Dilalla (Washington, DC: American Psychological Association, 2004): 161–89.

23. Michelle Alexander, *The New Jim Crow: Mass Incarceration in the Age of Colorblindness* (New York: New Press, 2012), 237.

24. This study is cited by Nancy Rubin, a teacher at Berkeley High School in her book *Ask Me If I Care: Voices from an American High School* (Berkeley: Ten Speed Press, 1994).

25. For a discussion of violence in marijuana farming, see Ray Raphael, "Green Gold and the American Way," in *West of Eden: Communes and Utopia in Northern California,* ed. Iain Boal et al. (Oakland, CA: P. M. Press, 2012), 187–98.

CHAPTER 8
RECONCILIATIONS

1. My challenges in dealing with Marco's adult addiction are little different from those of parents with addicted biological children. Some of the books on personal experiences with addiction that I found most useful are David Sheff, *Beautiful Boy: A Father's Journey through His Son's Addiction* (New York: Houghton, Mifflin, Harcourt, 2009); William Cope Moyers and Katherine Ketcham, *Broken: My Story of Addiction and Redemption* (New York, Penguin Books, 2007); Beverly Conyers, *Addict in the Family: Stories of Loss, Hope, and Recovery* (Center City, MN: Hazelden, 2003).

2. Throughout this book, the reader will have noticed my preoccupation with race. Because of continued racial segregation, I wanted to combat this politically and personally to make Marco feel less like an outsider. Of course, social class and educational achievement were also issues.

3. My birth cohort was much smaller than that of the baby boomers, who were born after the war. That meant there was less competition for university places and for jobs. By the time (1969) I completed my Ph.D., universities were expanding and there were many positions available, especially in the social sciences. I started teaching full-time in 1966 since there was such a need for college teachers. I always wanted to live in the San Francisco Bay Area and I could get a job here in the early 1970s long before the tech revolution raised the price of housing. Everything was much less expensive in my twenties and thirties than it is today. Neither I nor anyone I knew had college debt. Rents were low as were the price of houses. The house I bought in 1985, now more than thirty years later, is worth ten times what I paid for it. I've refinanced a number of times to live beyond my modest salary. Even my adoption expenses in 1981 were much less than they are today.

CHAPTER 9
EXTENDING FAMILY

1. Harold Grotevant et al., "Many Faces of Openness in Adoption: Perspective of Adopted Adolescents and Their Parents," *Adoption Quarterly* 10, nos.3–4 (2007); David Brodzinsky, "Reconceptualizing Openness in Adoption: Implications for Theory, Re-

search, and Practice," *Psychological Issues in Adoption: Research and Practice,* ed. David Brodzinsky and Jesus Palacios (Westport, CT: Praeger Press, 2005), chap. 7.

2. See Blaine Harden, "Born on the Bayou and Barely Feeling Any Urge to Roam," *New York Times,* September 30, 2002, page A8.

3. See Spike Lee's documentary on New Orleans after Katrina, *When the Levees Broke: A Requiem in Four Parts,* available on YouTube and Netflix. Also the documentary, *Faubourg Tremé: The Untold Story of Black New Orleans,* available on Amazon Prime and probably other places.

4. Since I have changed the name of the town for privacy reasons, I will not cite the newspaper report.

5. Chapter 6, "Adoption," in Andrew Solomon, *New Family Values.* Amazon Audible Books, 2018.

6. I have profiled some of these families on my blog on *Psychology Today* titled Adoption Diaries: Adult Relations in Adoption. www.psychologytoday.com/us/blog/adoption-diaries.

APPENDIX:
IMPLICATIONS FOR ADOPTION THEORY, PRACTICE, AND RESEARCH

1. Recently writers and journalist have exposed a number of corrupting practices. Kathryn Joyce, in *The Child Catchers: Rescue, Trafficking, and the New Gospel of Adoption* (New York: Public Affairs Press, 2013), investigates how religious organizations in the U.S. use adoption in an attempt to proselytize their views and she also documents trafficking in overseas adoption. A five-part series by Megan Twohey, "Americans Use the Internet to Abandon Children Adopted from Overseas" (published by Reuters in 2013, http://www.reuters.com/investigates/adoption/article), documents the use of the internet in elaborate rehoming schemes to transfer adoptees illegally within the U.S. Other personal exposures of child trafficking in overseas adoption are Elizabeth Larsen, "Did I Steal My Daughter?" *Mother Jones* (November 2007), and David M. Smolin, "The Aftermath of Unethical International Adoptions," *Point of View: The Newsletter for Adoptive Families with Children of Color, PACT* (Fall 2014).

2. The new field of Critical Adoption Studies encourages critiques of current adoption theory and discussions about reform of adoption practice. See Special Issue on Critical Adoption Studies in the journal *Adoption and Culture* 6, no. 1 (2018).

3. Sharon Vendivere, Karen Malm, and Laura Radel, *Adoption USA: A Chart Book Based on the 2007 National Survey of Adoptive Parents* (Washington, DC: U.S. Department of Health and Human Services, 2009), 33–35.

4. Michael Rutter, a British psychiatrist who uses behavioral genetics, did research on children from Romanian orphanages adopted in Britain. In the following article, he surveys results from a number of other researchers too: "Adverse Pre-adoption Experiences

and Psychological Outcomes," in *Psychological Issues in Adoption: Research and Practice*, ed. David M. Brodzinsky and Jesus Palacios (Westport, CT: Praeger, 2005), 67–92.

5. There are many books and articles on the history of adoption. Here are my favorites and those that have influenced my current analysis: Barbara Melosh, *Strangers and Kin: The American Way of Adoption* (Cambridge, MA: Harvard University Press, 2002); Judith S. Modell, *Kinship with Strangers: Adoption and Interpretation of Kinship in American Culture* (Berkeley: University of California Press, 1994); Julie Berebitsky, *Like Our Very Own: Adoption and the Changing Culture of Motherhood, 1851–1950* (Lawrence: University of Kansas Press, 2000); Herman, *Kinship by Design*.

6. Deborah H. Siegel and Susan Livingston Smith, *Openness in Adoption: From Secrecy and Stigma to Knowledge and Connections* (New York: Evan B. Donaldson Adoption Institute, 2012).

7. Christine Jones, "Openness in Adoption: Challenging the Narrative of Historical Progress," *Child and Family Social Work* (2013).

8. Harold Grotevant et al., "Many Faces of Openness in Adoption: Perspective of Adopted Adolescents and Their Parents," *Adoption Quarterly* 10, nos. 3–4 (2007); David Brodzinsky, "Reconceptualizing Openness in Adoption: Implications for Theory, Research, and Practice," in *Psychological Issues in Adoption: Research and Practice*, ed. David Brodzinsky and Jesus Palacios (Westport, CT: Praeger Press, 2005), chap. 7.

9. Harold Grotevant and Ruth G. McCoy, "Growing Up Adopted: Birth Parent Contact and Developmental Outcome," in *Experience and Development*, ed. Kathleen McCartney and Richard Weinberg (New York: Psychology Press, 2009), chap. 11.

10. One small qualitative study interviewed adoptive parents about their experiences with open adoption over eighteen years. The research concluded that open adoption provided information that adoptive parents found useful, including "knowing about genetic vulnerabilities." Deborah H. Siegel, "Open Adoption: Adoptive Parents' Reactions Two Decades Later," *Social Work* 58, no. 1 (January 2013): 43–53.

11. Thomas Bouchard et al., *Science* 250 (1990): 223–28.

12. Fiona Trotter and Laura Carin, "Open Borders: Openness in International Adoptions," *Focus on Adoption* 23 (Fall 2015); Jill Hodges, "Building Birth Family Connections beyond Borders," *Point of View: The Newsletter for Adoptive Families with Children of Color, PACT* (Fall 2014); Pamela Kruger, "To Search or Not to Search," in *A Love Like No Other: Stories from Adoptive Parents*, ed. Pamela Kruger and Jill Smolowe (New York: Riverhead Books, 2006), 22–33.

13. There is some evidence that some offspring decrease or cease using when they see the harm done to the parents. See Carol A. Prescott et al., "Challenges in Genetic Studies of the Etiology of Substance Use and Substance Use Disorders," *Behavior Genetics* 36 (2006): 473–82: 476.

14. Reunion in preadolescence or early adolescence is advocated by adoption educator and counselor Joyce Maguire Pavao in *The Family of Adoption*, rev. ed. (Boston: Beacon Press, 2005).

15. David Reiss, Leslie Leve, et al., "Understanding Links between Birth Parents and the Child They Have Placed for Adoption: Clues for Assisting Adoptive Families and Reducing Genetic Risk," in *International Advances in Adoption Research for Practice*, ed. Gretchen Miller Wrobel and Elspeth Neil (New York: Wiley-Blackwell, 2009), chap. 6, 131.

16. Aaron Panofsky, "Behavior Genetics and the Prospect of a 'Personalized Social Policy,'" *Policy and Society* 28 (2009): 328.

17. Richard P. Barth, "Outcomes of Adoption and What They Tell Us about Designing Adoption Services," *Adoption Quarterly* 6, no. 1 (2002): 58.

18. Aaron Panofsky, *Misbehaving Science: Controversy and the Development of Behavior Genetics* (Chicago: University of Chicago Press, 2014), 166.

19. One example is a Penn State behavioral geneticist, Jenae Neiderhiser, herself an adoptee. She has been presenting papers at adoption conferences (where I met her) and has published in *Adoption Quarterly*. Her work was profiled in the magazine *Adoptive Families*. See Rebecca Klein, "It's in Their Genes," *Adoptive Families* (February 2009).

20. I have not dealt with the design of twin research. An excellent overview is presented by Kenneth S. Kendler and Carol A. Prescott, *Genes, Environment and Psychopathology* (New York: Guilford Press, 2006).

21. Panofsky, *Misbehaving Science*.

22. ASHG/ ACMG, "Genetic Testing in Adoption," *American Journal of Human Genetics* 66 (2000): 761–67.

23. A birth mother's account of her open adoption with lots of contact for the first twelve years of her son's life is told by Amy Seek in *God and Jetfire: Confessions of a Birth Mother* (New York: Farrar, Straus & Giroux, 2015). Frank Ligtvoyt, a white adoptive cofather of two African American children, writes blogs on *Huffington Post* and other places about his positive experiences with open adoption. See "Real Parents in Adoption: A Paradigm Shift," http://www.huffingtonpost.com/frank-ligtvoet. I have interviewed several adoptive parents creating such kin relations and written about them on my blog, *Adoption Diaries: Adult Relationships in Adoption*, on Psychology Today website: https://www.psychologytoday.com/us/blog/adoption-diaries

24. See Andrew Solomon, *New Family Values*, Amazon Audible Books, 2018; and his book, *Who Rocks the Cradle*, forthcoming from Scribners.

25. See Rosanna Hertz and Margaret K. Nelson, *Random Families: Genetic Strangers, Sperm Donor Siblings, and the Creation of New Kin* (New York, Oxford University Press, 2019); and Dani Shapiro, *Inheritance: A Memoir of Genealogy, Paternity, and Love* (New York: Knopf, 2019).

INDEX

Bechtle, Robert, 43–44

behavioral genetics: attention span persistence scale, 32; on children building their own worlds, 89; on cognitive abilities, 29–30; on divorce, 61–62; on educational achievement, 31–32; ethical concerns, 170–71; field of, 1–2, 28–29; general findings of, 29, 35; implications for adoption, 168–70; on increased genetic impact over time, 96, 162; on life outcomes, 34; on nonshared environment, 186n22; on parenting and family environment, 94–96; parents' genes, impact of, 94; on peer affiliation, 105–9; on personality traits, 33–34; on reading ability, 31; sibling studies, 170; twin studies, 28, 47, 170, 185n20; on verbal ability, 30–31. *See also* environmental factors

Behavioral Genetics (Plomin), 96

Berkeley, CA: about, 54–55, 180n1; communal household in, 8–9, 54–61; communes in, 180n6; house in, 80–81, 187n3; outpatient rehab program, 120–21, 132. *See also* schools

Berkeley Alternative High School, 104–5

Berkeley High School, 103–4

Big Brother program, 84

biological family. *See* birth families of Marco; extended family, incorporating adoptive and biological kin

birth families of Marco: aunt, 38–42, 141–43; birth father, 12–13, 23–26, 42–43, 51–53, 138–43, 148–51, 157; Marco's reflections on, 153–55; mother, 16–20, 23–27, 37–38, 44–46, 50–52, 141–44, 157; reunion with, 20, 22–25; search for birth parents, 19–20, 179n22; sisters, 147–50, 151, 155; step-grandmother, 146; stepmother, 17–18, 51,

138–41, 146–47, 151; uncle, 14, 30–31, 39, 45, 144–47

Blank, Les, 17

blank slate theory, 28, 162. *See also* nature vs. nurture

Bouchard, Thomas, 164

Brodzinsky, David, 169, 175n6

Broyard, Bliss, 15

Cajuns, 16–17, 137–38

Caplan, Bryan, 185n20

The Child Catchers (Joyce), 188n1

child sexual assault prevention education, 69, 71, 182n16. *See also* sexual abuse

class biases, 24–25, 85, 176n8

Clean: Overcoming Addiction (Sheff), 46–47

cognitive abilities study, 29–30

Colorado Adoption Project (CAP), 29–31, 61–62, 176n11

communal household, 8–9, 54–61, 75

Cornell University, 93

Creoles, 15–17, 137–38

Creoles of Louisiana (Istre), 176n8

Critical Adoption Studies, 188n2

cultural roots, Creole and Cajun, 15–16, 137–38

dance, 87–88

delinquencies and addiction, 21

DePaulo, Bella, 61

Diagnostic and Statistical Manual of Mental Disorders (DSM-5), 49

divorce, 61–62, 180n15

dopamine, 48

drug dealing, 103, 104, 110, 118

drug use. *See* addiction and substance abuse; marijuana

DUI, 126

DuPont, Robert L., 46

hunting, 14, 145

Hurricane Katrina, 137

impulsivity, 49, 50

interests, personal, 15, 41, 44, 81–82, 92–94

international adoptions, 162, 165

Iowa adoption study, 47–48, 49, 62, 78, 183n18

Istre, Elista, 176n8

jails, 21, 25, 41–42, 45, 117–18, 127, 144, 155

jazz dance, 87–88

Jensen, Arthur, 28

Joyce, Kathryn, 188n1

Kinship by Design (Herman), 189n5

Kinship with Strangers (Modell), 189n5

Lake Charles, LA, 143–44

A Lesson before Dying (Gaines), 137

Leve, Leslie, 166–67, 184n13

life outcomes study, 34

Ligtvoyt, Frank, 190n23

Losing Isaiah (film), 153

Louisiana: author's solo visit to, 137–46; hometown of birth parents, 138, 141–43; jail in, 41–42; Lake Charles, 143–44; Marco's earliest questions and learning about, 15–16; Marco's first trip to, 22–25; Marco's long stay in, 26–27, 37–46, 50–53; New Orleans, 17, 138–40, 157; prisons in, 176n9; swamp tour, 137; workshop on Creole and Cajun culture, 137–38

lying, 110

Maccoby, Eleanor, 94

marijuana: Berkeley culture of tolerance and, 110–11; early exposure, impact of, 111–12; first use of, 89, 110; genetic factors in use of, 112; grades and, 110; growing, 118–19, 129; potency of marijuana in 1990s vs. 1960s, 111; prefrontal cortex development and, 112; rates of addiction, 111, 186n11; selling, 20, 103, 104, 107, 110, 113, 130. *See also* addiction and substance abuse; rehab programs

masculinity, 116, 117

Melosh, Barbara, 189n5

memory, childhood, 77, 79

Mengele, Gregor, 28

methylation, 183n20

Miller, Timothy, 180n6

Minnesota longitudinal adoption study, 47, 62, 179n10

misdemeanor charges, 20–21

Mitchell Canyon, 124–25

mixed-race children: as adoptees, 9–10; in Berkeley neighborhood, 81, 102; in Louisiana, 19, 24, 146, 153; in schools, 71, 84

mixed-race population, in Berkeley and Oakland, 54

Modell, Judith S., 189n5

Modern Families (Golombok), 183n2

Mother's Day, 43–44, 124–25

multiracial environment, 83

National Association of Black Social Workers, 10

nature vs. nurture: blank slate theory, 28, 162; impact of nature, 161–62; as interrelated, 28–29; primacy of nurture theory ("nurture is everything"), 7, 113–14, 162. *See also* behavioral genetics

Neiderhiser, Jenae, 177n12, 179n20, 190n19

Nelson, Margaret, 190n25

CPSIA information can be obtained
at www.ICGtesting.com
Printed in the USA
LVHW021946180220
647337LV00006B/837

9 780807 173107